Centre for Innovation in Mathematics Teaching
University of Exeter

DISCRETE
MATHEMATICS

Writers   David Burghes
             John Deft
             Nigel Green
             Nigel Price

Editor   Nigel Price

Assistant   Ann Ault
Editors   Peter Gwilliam

Heinemann Educational

Heinemann Educational

a division of Heinemann Educational Books Ltd.

Halley Court, Jordan Hill, Oxford OX2 8EJ

OXFORD LONDON EDINBURGH
MADRID ATHENS BOLOGNA PARIS
MELBOURNE SYDNEY AUCKLAND SINGAPORE
TOKYO IBADAN NAIROBI HARARE
GABORONE PORTSMOUTH NH (USA)

ISBN 0 435 51608 6

First Published 1992

Second edition published 1994

94 95 96 97 98 6 5 4 3 2

© CIMT, 1992

Typeset by ISCA Press, CIMT, University of Exeter

Printed in Great Britain by The Bath Press, Avon

DISCRETE MATHEMATICS

This is one of the texts which has been written to support the AEB Mathematics syllabus for A and AS level awards first available in Summer 1996.

The text, which covers the Discrete Mathematics Module 4 , is a development of the Decision Mathematics Core text written for the AEB Coursework syllabus.

The development of these texts has been coordinated at the

Centre for Innovation in Mathematics Teaching

at Exeter University in association with Heinemann and AEB. The overall development of these texts has been directed by David Burghes and coordinated by Nigel Price.

Enquiries regarding this project and further details of the work of the Centre should be addressed to

Margaret Roddick
CIMT
School of Education
University of Exeter
Heavitree Road
EXETER EX1 2LU.

CONTENTS

PREFACE

The latter half of the 20th century has seen rapid advances in the development of suitable techniques for solving decision-making problems. A catalyst for these advances has been the revolution in computing. Even the smallest business is able to obtain, relatively cheaply, significant computer power. This has led to a rapid increase of research into **discrete** or **finite** mathematics.

The vast majority of mathematical developments over the past 300 years since Newton have concentrated on advances in our understanding and application of **continuous** mathematics, and it is only recently that parallel advances in **discrete** mathematics have been sought. Much, although not all, of the mathematics in this text is based on developments this century which are still continuing.

The aim of this text is to give a comprehensive treatment of the major topics in discrete mathematics, emphasising their applicability to problems in a highly technological world.

This text has been produced for students and includes examples, activities and exercises. It should be noted that the activities are **not** optional but are an important part of the learning philosophy in which you are expected to take a very active part. The text integrates

- **Exposition**   in which the concept is explained;
- **Examples**   which show how the techniques are used;
- **Activities**   which either introduce new concepts or reinforce techniques;
- **Discussion Points**   which are essentially 'stop and think' points, where discussion with other students and teachers will be helpful;
- **Exercises**   at the end of most sections in order to provide further practice;
- **Miscellaneous Exercises**   at the end of each chapter which provide opportunities for reinforcement of the main points of the chapter.

Discussion points are written in a special typeface as illustrated here.

Note that answers to the exercises are given at the back of the book. You are expected to have a calculator available throughout your study of this text and occasionally to have access to a computer.

Some of the sections, exercises and questions are marked with an asterisk (*). This means that they are either **not** central to the development of the topics in this text and can be omitted without causing problems, or they are regarded as particularly challenging.

This text is one of a series of texts written specially for the new AEB Mathematics syllabus for A and AS level coursework. The framework is shown opposite. Essentially each module corresponds to an AS level syllabus and two suitable modules provide the syllabus for an A level award. Optional coursework is available for students taking any of the three applied modules

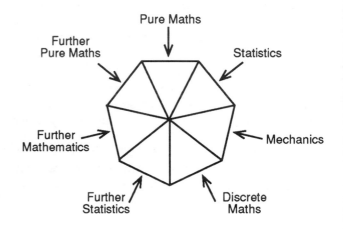

Mechanics, Statistics and Discrete Mathematics.

Full details of the scheme are available from
 AEB, Stag Hill House, Guildford GU2 5XJ.

We hope that you enjoy working through the book. We would be very grateful for comments, criticisms and notification of any errors. These should be sent to

Margaret Roddick
CIMT
School of Education
University of Exeter
EXETER EX1 2LU

from whom full details of other CIMT publications and courses for both teachers and students are available.

ACKNOWLEDGEMENTS

This text has been developed from an earlier version, published specifically for the CIMT/AEB Mathematics syllabus and based on the philosophy of the Wessex Project. I am particularly grateful to the original authors and to Bob Rainbow (Wessex) and John Commerford (AEB) for their help and encouragement in the development of the original resources.

This revised version and associated texts have been written for the new AEB Mathematics syllabus and assessment, which will be examined for the first time in Summer 1996. I am grateful for the continued support from AEB through its mathematics officer, Jackie Bawden, and to the staff at Heinemann, particularly Philip Ellaway.

Finally, I am indebted to the staff at CIMT who work with dedication and good humour despite the pressure which I continually put upon them. In particular, I am grateful to Nigel Price for the revisions and editing of the text, to Ann Ault and Peter Gwilliam for checking the mathematics and to Ann Tylisczuk and Liz Holland for producing camera-ready copy.

David Burghes
(Series Editor)
February 1994

1 GRAPHS

Objectives

After studying this chapter you should

- be able to use the language of graph theory;
- understand the concept of isomorphism;
- be able to search and count systematically;
- be able to apply graph methods to simple problems.

1.0   Introduction

This chapter introduces the language and basic theory of graphs.
These are not graphs drawn on squared paper, such as you met
during your GCSE course, but merely sets of points joined by lines.
You do not need any previous mathematical knowledge to study
this chapter, other than an ability to count and to do very simple
arithmetic.

Although graph theory was first explored more than two hundred
years ago, it was thought of as little more than a game for
mathematicians and was not really taken seriously until the late
twentieth century. The growth in computer power, however, led to
the realisation that graph theory can be applied to a wide range of
industrial and commercial management problems of considerable
economic importance.

Some of the applications of graph theory are studied in later
chapters of this book. Chapter 2, for example, looks at several
different problems involving the planning of 'best' networks or
routes, while Chapter 6 considers the question of planarity (very
important in designing microchips and other electronic circuits).
Chapter 7 deals with problems to do with the flow of vehicles
through a road system or oil through a pipe, and Chapters 12 and
13 show how to analyse a complex task and determine the most
efficient way in which it can be done. All these applications,
however, depend on an understanding of the basic principles of
graph theory.

1.1   The language of graphs

A **graph** is defined as consisting of a set of **vertices** (points) and a
set of **edges** (straight or curved lines; alternatively called arcs):
each edge joins one vertex to another, or starts and ends at the
same vertex.

The diagrams show three different graphs, representing respectively the major roads between four towns, the friendships among a group of students, and the molecular structure of acetic acid - the theory of graphs can be applied in many different ways.

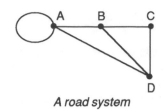

A road system

There are several things to note. One is that although nearly all the edges in these graphs have been drawn as straight lines, this is purely a matter of convenience. Curved lines would have done just as well, because what matters is which vertices are joined, not the shape of the line joining them. Second, each edge joins only two vertices, so that ABC in the first graph is two edges (AB and BC) rather than one long one. Third, the crossing in the middle of the second diagram is not a vertex of the graph; the only points counted as vertices are the ones identified as such at the start.

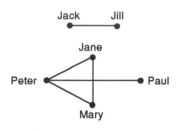

Friendships

The **degree** of a vertex is defined as the number of edges which start or finish at that vertex - an edge which starts and finishes at the same vertex (in other words, a **loop** such as the one at A in the first graph) is counted twice. So, for example, the degree of the vertex A in the first graph is 4, and the degree of the vertex 'Peter' in the second graph is 3. In the third case, the degree of each vertex corresponds to the valency of the atom.

There is actually something a little unusual in the third graph - two edges joining the same two vertices. A multiple edge of this kind can be of great importance in some situations: the difference between saturated and unsaturated fats in a healthy diet, for example, is largely a matter of multiple edges in their molecular structure. In other cases, however, such as the second graph here, a double or triple edge would be meaningless. A graph with no loops and no multiple edges is called a **simple** graph.

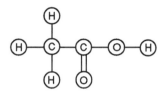

Acetic Acid CH_3COOH

There is an oddity in the second graph too. Jack and Jill are friends with one another but with no one else, so that the graph 'falls apart' into two quite separate pieces. Such a graph is said to be **disconnected**. A **connected** graph is one in which every vertex is linked (by a single edge or a sequence of edges) to every other. If every vertex is linked to every other by a single edge, a simple graph is said to be **complete**.

A **subgraph** of a graph is another graph that can be seen within it; i.e. another graph consisting of some of the original vertices and edges. For example, the graph consisting of vertices 'Jane', 'Mary' and 'Peter' and edges from 'Jane' to 'Mary' and from 'Mary' to 'Peter' is a subgraph of the friendship graph above.

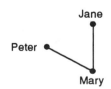

Exercise 1A

Note The answers to these questions will be used in later sections, and should be kept safely until then.

1. For each of the graphs shown below, write down
   (i) its number of vertices,
   (ii) its number of edges,
   (iii) the degree of each vertex.

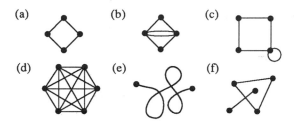

(a)   (b)   (c)

(d)   (e)   (f)

2. Say which (if any) of the graphs in Question 1 are
   (i) simple (ii) connected and/or (iii) complete.

3. Draw graphs to fit the following descriptions:

   (a) The vertices are A, B, C and D; the edges join AB, BC, CD, AD and BD.

   (b) The vertices are P, Q, R, S and T, and there are edges joining PQ, PR, PS and PT.

   (c) The graph has vertices W, X, Y and Z and edges XY, YZ , YZ, ZX and XX.

   (d) The graph has five vertices, each joined by a single edge to every other vertex.

   (e) The graph is a simple connected graph with four vertices and three edges.

1.2   Isomorphism

Look at your answers to Question 3 from Exercise 1A, and compare them with those of other students. You will probably find that some of the drawings look different from others and yet fit the descriptions equally well.

Two graphs which look different, but both of which are correctly drawn from a full description are said to be **isomorphic** - the word comes from Greek words meaning 'the same shape'. Isomorphism is a very important and powerful idea in advanced mathematics - it crops up in many different places - but at heart it is really very simple.

For example, the two graphs shown in the upper diagram each match the full description in Question 3(a) and so are isomorphic to one another. The graphs in the lower diagram each match the description in Question 3(e), but these are not isomorphic. The description did not say which vertices were to be joined by the edges, and the two graphs have joined the vertices differently.

If you are to say that two graphs are isomorphic, there must be a way of labelling or relabelling one or both of them so that the number of edges joining A to B in the first is equal to the number of edges joining A to B in the second, and so on through all possible pairs of vertices. In the upper diagram this is clearly possible: the labelling already on the graphs satisfies this condition, and indeed many people would say that the two graphs are more than isomorphic - they are identical. In the lower diagram, however, no such labelling can ever be found. The second graph has one vertex which is joined to three others, and no labelling of the first graph can ever match this.

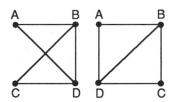

Two possible answers to Question 3 (a)

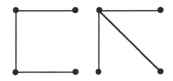

Two possible answers to Question 3 (e)

Testing for isomorphism

If you can match the labels it certainly shows that two graphs are isomorphic, but suppose you cannot. What does that show? It might mean that the graphs are not isomorphic, or it might simply be that you have not yet tried the right labelling combination.

How can you know if the two graphs are isomorphic?

The clue is in the argument that has already been given. If one graph has a vertex of degree three, and the other does not, then no matching can ever be found and the graphs are not isomorphic. This idea can be extended to provide a partial test: a **necessary** condition for two graphs to be isomorphic is that the two graphs have the same number of vertices of degree 0, the same number of vertices of degree 1, and so on. If this condition is not satisfied the graphs are certainly not isomorphic. But it is not a **sufficient** condition; in other words, if the condition is satisfied you still do not know whether or not the graphs are isomorphic and you must go on looking for a match.

Exercise 1B

Look at the graphs below, and say which of them are isomorphic to which others.

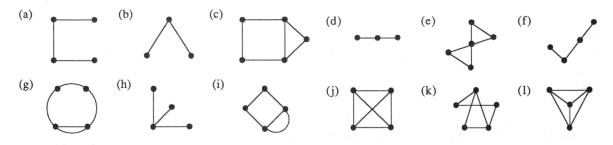

Counting graphs

You may be wondering by now how many different simple graphs can be drawn with just a few vertices. With only one vertex (and no loops allowed) there is clearly only one such graph - the one with no edges.

With two vertices there are two possibilities: there is one graph with no edges and one with one edge, making two possible graphs altogether. You are counting simple graphs, remember, so multiple edges are excluded.

With three vertices there are four possibilities: one each with no edges, one edge, two edges and three edges respectively. Any other simple graph on three vertices must be isomorphic to one of these.

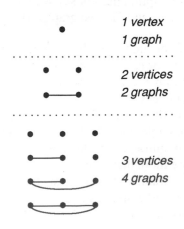

1 vertex
1 graph

2 vertices
2 graphs

3 vertices
4 graphs

*Activity 1   Simple graphs

These results can be put into a table:

Vertices	1	2	3	4	5	6	...
Graphs	1	2	4	...	...	...	...

Try to predict the number of different simple graphs that can be drawn on four vertices, and then check your prediction by drawing them. It is worth doing this in discussion with another student to ensure that you do not leave out any possible graphs nor include two that are actually isomorphic.

When you have got a firm result for four vertices (and corrected your prediction if necessary), try to extend your prediction to five and/or six vertices.

Activity 2   Handshakes

At the beginning of the lesson, greet some of the other members of your group by shaking hands with them. You don't have to shake hands with everyone, and you can shake hands with the same person more than once if you like, but you must keep count of how many handshakes you take part in.

At the end, some members of the group will have been involved in an odd number of handshakes, and others in an even number, so consider this bet: if the number of people involved in an odd number of handshakes is odd, your teacher lets you off homework for a week, but if it is even you get a double dose - does that seem fair?

You may guess that this is not a good bet at all from your point of view - not unless you like doing maths homework, that is! In fact you can never win, because the number of people who shake hands an odd number of times is always even.

Look at your answers to Question 1 in Exercise 1A. Any handshaking situation can be represented by a graph, with people as vertices and handshakes as edges; it may have multiple edges, but not loops. For each graph, find the sum of the degrees of the vertices, and compare it with your other data.

The handshake lemma

You can see at once that the sum of the degrees of the vertices is always twice the number of edges. This is known as the handshake lemma - a lemma is a mini-theorem - and is easy to prove. The degree of each vertex is the number of edge-ends at that vertex, and since each edge has two ends, the number of edge-ends (and hence the total of the vertex degrees) must be twice the number of edges.

This lemma leads quite easily to the unwinnable bet. If the total of individual handshakes is twice the number of handshakes, as the lemma requires, it is certainly an even number. Some members of the class shook hands an even number of times, and the total of any number of even numbers is even. So the total for the rest must be even as well, and since they each shook hands an odd number of times, this can happen only if there are an even number of people. So the number of people involved an an odd number of handshakes must always be even.

The handshake lemma may seem trivial, but it has some quite important consequences and comes up again in Chapter 6.

Activity 3

Try the handshaking exercise again, and this time keep count not of the number of handshakes, but of the number of people with whom you shake hands (once or more times makes no difference). What are the chances that at the end there will be two people who have shaken hands with the same number of others?

The pigeonhole principle is considered in more depth in Chapter 3.

It will perhaps not surprise you to learn that such a coincidence is certain to happen. The proof of this depends on a simple but important principle known as the pigeonhole principle. If n objects have to be put into m pigeonholes, where $n > m$, it says that there must be at least one pigeonhole with more than one object in it. Like the handshake lemma, the pigeonhole principle seems obvious but has a number of uses.

For example, suppose there are nine people in the room: each must have shaken hands with 0, 1, 2, 3, 4, 5, 6, 7 or 8 others. Of course, if anyone has shaken hands with 8 others - that is, with everyone else - then there cannot be anyone who has shaken hands with 0 others, and vice versa. So among the nine people there are at most eight different scores and the pigeonhole principle says that at least two people must therefore have the same score. You can apply the same argument to any number of people more than one.

1.3 Walks, trails and paths

If you have read any other books about graph theory, you may find this next section rather confusing. Graph theory is a relatively new branch of mathematics, and as yet there is no universal agreement as to the meanings to be given to certain terms. The consequence is that what is called a trail here might be called a walk in another book and a path in a third - the ideas are common but the words are different. The definitions to be used in this book are as follows:

A **walk** is a sequence of edges of a graph such that the second vertex of each edge (except for the last edge) is the first vertex of the next edge. For example, the sequence CD, DA, AB, BD, DA defines a walk (which might be called a walk from C to A) in the graph shown in the diagram. A walk can be the trivial one with no edges at all!

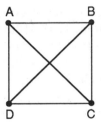

A **trail** is a walk such that no edge is included (in either direction) more than once in the sequence. The walk above is not a trail because the edge DA occurs twice, but CD, DA, AB, BC, CA is a trail from C to A.

A **path** is a trail such that no vertex is visited more than once (except that the first vertex may also be the last); the trail above is not a path because both A and C are visited more than once, but CD, DA, AB is a path from C to B.

Exercise 1C

1. Referring to the graph in the diagram below, list

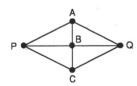

(a) all the paths from P to Q;

(b) at least three trails from P to Q which are not paths;

(c) at least three walks from P to Q which are not trails;

(d) all the paths which start and finish at P.

2. Which (if any) of the shapes below can you draw completely without lifting your pencil from the paper or going over any line twice?

(If you invent appropriate vertices and imagine them as graphs, then you are looking for a trail which includes all the edges)

(a)   (b)   (c)

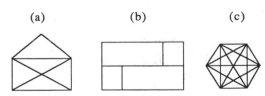

1.4   Cycles and Eulerian trails

Cycles

Puzzles involving trails and paths have been popular for many years, and you may well have seen some or all of the graphs above in books of recreational mathematics. Of particular interest are walks, etc which start and finish at the same place; a walk, trail or path which finishes at its starting point is said to be **closed**, and a closed path with one or more edges is called a **cycle**.

Modern graph theory effectively began with a problem concerning a closed trail. In the 18th century the citizens of the Prussian city of Königsberg (now called Kaliningrad) used to occupy their Sunday afternoons in going for walks. The city stood on the River Pregel and had seven bridges, arranged as shown in the diagram. The citizens' aim was to find a route that would take them just once over each bridge and home again.

Königsberg bridges

Activity 4   Königsberg bridges

Try to find a route crossing each bridge just once and returning to the starting point.

If you failed, don't.worry - so did the people of Königsberg! They began to realise that such a route was impossible, but it was some years before the great Swiss mathematician *Leonhard Euler* (1707-83) proved that this was indeed so. The modern proof, developed from Euler's, is very simple once the bridges are represented by edges of a graph.

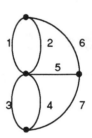

Königsberg bridges in graph form

How can you be sure that there is no closed trail using all the edges of this graph?

Eulerian trails

The four vertices of the graph have degrees 3, 3, 3 and 5 respectively - all odd numbers. Any closed trail, on the other hand, goes into a vertex and out of it again, thus adding 2 to its degree on each visit. A closed trail using all the edges cannot exist, therefore, unless every vertex has even degree. (If there are just two vertices with odd degree, they could be the start and finish of a non-closed trail using all the edges.)   In fact the opposite is also known: if a connected graph has every vertex of even degree then there does exist a closed trail using all the edges (and if there are just two vertices of odd degree then there is a non-closed trail using all the edges).

As a mark of respect for Euler's work in this area, a trail which includes every edge of a graph is called an **Eulerian trail**. If the trail is closed, the graph itself is said to be **Eulerian**; a semi-Eulerian graph is one that has a non-closed trail including every edge.

Exercise 1D

By considering the degree of each vertex, determine whether each of the graphs shown opposite is Eulerian, semi-Eulerian, or neither. In the case of Eulerian and semi-Eulerian graphs, find an Eulerian trail.

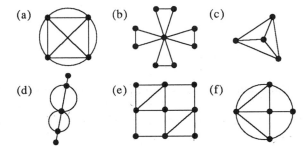

1.5   Hamiltonian cycles

Sir William Rowan Hamilton (1805-65) was a nineteenth-century Irish mathematician who invented in his spare time a game called the **Icosian Game**, based on the vertices of an icosahedron. The idea was essentially simple: given the first five vertices, the player had to find a route that would pass through the remaining fifteen and return to the start without using any vertex twice.

Activity 5   Icosian game

The diagram shows a graph representing a dodecahedron. Try to find such a route - a closed path, to use the modern phrase - beginning with ABCIN in that order.

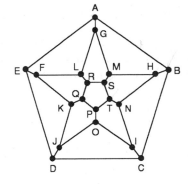

A closed path that passes through every vertex of a graph is called a **Hamiltonian cycle**, and a graph in which a Hamiltonian cycle exists is said to be **Hamiltonian**. The dodecahedron is a Hamiltonian graph, and there are actually two Hamiltonian cycles beginning with the five vertices given:

ABCINHMSTPOJDEFKQRLGA

and   ABCINHMGLFKQRSTPOJDEA.

Distinguishing Hamiltonian from non-Hamiltonian graphs is not easy, and there is no simple test corresponding to the even-degree test for Eulerian graphs.

* Exercise 1E

Decide by trial and error whether or not each of the graphs shown below is Hamiltonian.

(a)   (b)   (c)   (d)   (e)   (f)

1.6 Trees

A connected graph in which there are no cycles is called a **tree**.

Look at the graphs below and decide which of them are trees.

(a)   (b)   (c)   (d)

(e)  (f)   (g) (h)

Activity 6

Look again at the graphs you have identified as trees, and count their vertices and their edges. Can you state a general theorem connecting the numbers of vertices and edges for trees? If so, can you prove it?

It is fairly easy to guess from the examples that the number of edges of a tree is always one less than the number of vertices. The proof too is straightforward: if the tree is built up one vertex at a time, starting with one vertex and no edges, each new vertex needs exactly one edge to join it to the body of the tree.

Trees of this kind occur quite often in real life - a biology book may include a 'tree' showing how all living creatures are ultimately descended from the same primitive life forms; a geography text may include a diagram of the entire Amazon river system; and you may find in a history book a diagram of the Kings and Queens of England, although a certain amount of intermarriage prevents this from being truly a tree as defined above.

Hierarchies

Trees are also commonly used to represent hierarchical organisations. The first diagram below shows part of the management structure of a college, for example, while the second is an extract from a computer's hard disk directory.

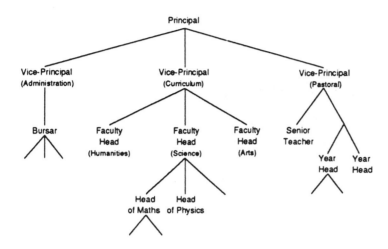

A management Tree

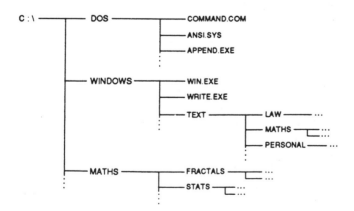

A computer directory tree

Game strategies

Another application of trees is in setting out strategies for playing certain games. For example, the diagram shows the first few stages of a strategy tree for the first player in the game "Noughts and Crosses". You may like to try to complete it.

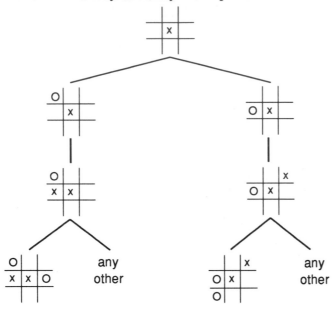

A strategy tree for 'Noughts and Crosses'

Counting trees

There is clearly only one tree with one vertex, one with two, and one with three, as shown in the diagram - any other is isomorphic to one of these. There are two non-isomorphic trees with four vertices, however, and these figures can be set out in a table:

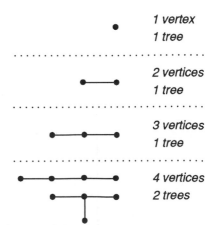

1 vertex
1 tree

2 vertices
1 tree

3 vertices
1 tree

4 vertices
2 trees

Vertices	1	2	3	4	5	6	7	...
Trees	1	1	1	2	...	...	...	...

*Activity 7   Counting trees

Draw all the different trees with five vertices, and all those with six, and try to predict from your results the number of seven-vertex trees. Check your prediction by drawing them.

*Organic chemistry

In fact there is no simple formula for unlabelled trees - it turns out to be much easier to count trees if their vertices are labelled - but even the few results in your table can be of use in identifying chemical compounds.

The group of chemicals known as alkanes have molecules made up of a carbon 'tree' surrounded by hydrogen: each hydrogen atom is bonded to one carbon, and each carbon atom is bonded to four other atoms, of either kind. The diagram shows a molecule of propane (the fuel used in some camping gas stoves), which has three carbon and eight hydrogen atoms and so has molecular formula $C_3 H_8$.

The molecule can actually be represented completely by its carbon tree, because once all the carbon atoms have been bonded in some formation the hydrogen atoms must go wherever there is a free bond. Now according to the table above there is only one possible carbon tree for CH_4 (methane), only one for $C_2 H_6$ (ethane), and only one for $C_3 H_8$ (propane), so each of these molecular formulae represents only one compound. But there are two distinct trees with four vertices, so the formula $C_4 H_{10}$ can represent either butane or isobutane, two different compounds with different properties.

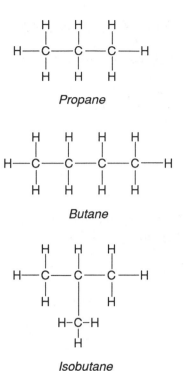

Propane

Butane

Isobutane
(or 2-methyl-propane)

* Exercise 1F

1. How many different compounds have the molecular formula $C_5 H_{12}$? (If you are studying A Level Chemistry, what are their names?)

2. How many different compounds have molecular formula $C_6 H_{14}$? Think carefully before you answer.

*1.7 Coloured cubes

You may have seen in the shops a puzzle consisting of four cubes with different colours or other designs on their sides. The aim of the puzzle is to stack the cubes in a tower so that each of the long faces shows four different colours or designs. A trial-and-error approach is very difficult, but the application of a little graph theory can lead directly to a solution.

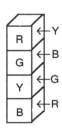

Example

Suppose that the four cubes are coloured as shown .

Cube 1		
red	*opposite*	yellow
green	*opposite*	yellow
blue	*opposite*	red

Cube 2		
red	*opposite*	red
green	*opposite*	blue
blue	*opposite*	yellow

Cube 3		
red	*opposite*	green
blue	*opposite*	green
blue	*opposite*	yellow

Cube 4		
red	*opposite*	yellow
green	*opposite*	blue
green	*opposite*	yellow

Transfer all this information to a graph as shown in the diagram, joining the vertices representing opposite colours by an edge numbered to show the cube to which it belongs.

From this graph, extract two disjoint subgraphs - that is, two subgraphs with no edges in common. Each subgraph must consist of four edges of the original graph, chosen in such a way that

(i)   the edges include one of each number, in any order,

(ii) each of the vertices R, Y, G, B has degree 2 in each subgraph.

Two subgraphs satisfying these conditions are shown in the lower diagram opposite.

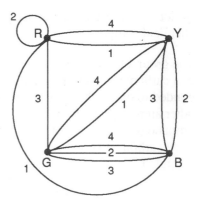

A graphical representation of the four cubes

The subgraphs now tell you how to stack the cubes:

	From first subgraph		From second subgraph	
	Front	**Back**	**Left**	**Right**
Cube 1	green	yellow	red	blue
Cube 2	red	red	blue	yellow
Cube 3	yellow	blue	green	red
Cube 4	blue	green	yellow	green

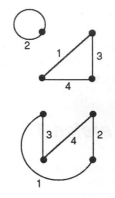

Two disjoint subgraphs

In this particular case there are two other solutions - there is a third subgraph that could have been chosen with either of the two above.

Activity 8

Find the third subgraph and interpret it in the same way as in the example.

Some patterns of cubes have only one solution, however, and others have no solution at all.

*Activity 9

Try to get hold of a commercially-made puzzle of this kind, or make your own, and then amaze your family and friends (and perhaps yourself!) by using mathematics to solve it in just a few minutes.

1.8 Miscellaneous Exercises

1. Draw two simple connected graphs, each with four vertices and four edges, which are not isomorphic.

2. If the vertices of a graph have degree 1, 2, 2, 2 and 3 respectively, how many edges has the graph? Draw two simple connected graphs, each with this vertex set, which are not isomorphic.

3. If P, Q, R, S and T are the vertices of a complete graph, list all the paths from S to T.

4. Determine whether each of the following graphs is Eulerian, semi-Eulerian or neither, and find an Eulerian trail if one exists.

(a) (b) (c)

5. You are given nine apparently identical coins, eight of which are genuine, the other being counterfeit and different in weight from the rest - either heavier or lighter, but you do not know which. You are also given a two-sided balance on which to compare the weights of coins or groups of coins. Draw a tree to show a strategy for identifying the counterfeit coin in no more than three weighings.

*6. How many different compounds have molecular formula C_7H_{16}?

*7. Find a Hamiltonian cycle on the graph shown in the diagram.

A four-dimensional cube in graph form

*8. Prove that among any group of six people, there are either three who all know one another or three who are mutual strangers.

*9. Given that there are 23 different unlabelled trees with eight vertices, draw as many of them as you can.

*10. A set of four coloured cubes has opposite faces coloured as follows:

Cube 1	R-B, R-Y, B-G;
Cube 2	R-B, Y-Y, Y-G;
Cube 3	R-Y, R-B, B-G;
Cube 4	R-G, G-G, B-Y;

Either find a solution to the four-cube problem or explain why such a solution is impossible.

11. Consider the following graphs G_1 and G_2:

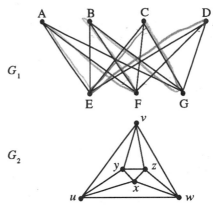

(a) Determine whether each of these graphs is Eulerian. In each case, either give an Eulerian trail or state why such a trail cannot exist.

(b) Determine whether each of these graphs is Hamiltonian. In each case, either give a Hamiltonian cycle or state why such a cycle cannot exist.

12. A simple graph G has five vertices, and each of those vertices has the same degree d.

(a) State the possible values of d.

(b) If G is connected, what are the possible values of d?

(c) If Eulerian, what are the possible values of d?

(AEB)

13. Consider the following graph of G.

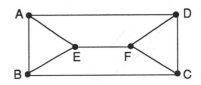

(a) Is G Eulerian? If so, write down an Eulerian trail.

(b) Is G Hamiltonian? If so, write down a Hamiltonian cycle.

2 TRAVEL PROBLEMS

Objectives

After studying this chapter you should

* be able to explain the shortest path, minimum connector, travelling salesman and Chinese postman problems and distinguish between them;

* understand the importance of algorithms in solving problems;

* be able to apply a given algorithm.

2.0   Introduction

In this chapter the ideas developed in Chapter 1 are applied to four important classes of problem. The work will make very little sense unless you have studied Chapter 1 already, but no other mathematical knowledge is required beyond some basic arithmetic. A simple calculator may be helpful for some of the work.

2.1   The shortest path problem

The graph opposite represents a group of towns, with the figures being the distances in miles between them. It may seem strange that the direct edge OK is longer than the two edges OH and HK added together, but in real life this is not unusual. You can probably think of examples in your own area where the signposted route from one place to another is longer than an 'indirect' route via a third place, known only to local people.

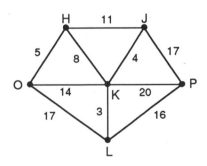

Activity 1   Shortest path

Find by trial and error the shortest route from O to P.

The diagram opposite shows a number of villages in a mountainous area and the time (in minutes) that it takes to walk between them. Once again you may notice that there are cases where two edges of a triangle together are 'shorter' than the third - this is perhaps because the third edge goes directly over the top of a mountain, while the first two go around the side.

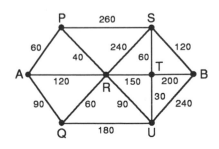

Activity 2   Shortest path

Find the quickest route from A to B.

You may think that there should be some logical and systematic way of finding a shortest route, without relying on a lucky guess. As the number of vertices increases it becomes more and more important to find such a method; and if the problem is to be turned over to a computer, as is usual when problems such as these arise in real life, it is not just important but essential.

What is actually needed is an **algorithm** - that is, a set of step-by-step instructions that can be applied automatically, without any need for personal judgement. A recipe in a cookery book is a good example of an algorithm, as is a well-written set of directions for finding someone's house. The 'Noughts and Crosses' strategy in Section 1.6 is another example, and any computer program depends on an algorithm of some kind.

Activity 3   Finding an algorithm

On your own or in discussion with another student, try to write an algorithm for solving a shortest path problem. That is, write a set of instructions so that someone who knows nothing about mathematics (except how to do simple arithmetic) can solve any such problem just by following your instructions. Test your algorithm on the examples on the previous page.

The shortest path algorithm

The usual algorithm sometimes known as Dijkstra's method is set out below; you may have come up with something similar yourself, or you may have found a different approach. Study this algorithm anyway, and then decide how you want to handle problems such as these in the future.

Consider two vertex sets: the set S, which contains only the start vertex to begin with, and the set T, which initially contains all the other vertices. As the algorithm proceeds, each vertex in turn will be labelled with a distance and transferred from T to S.

The algorithm runs as follows:

1. Label the start vertex with distance 0.

2. Consider all the edges joining a vertex in S to a vertex in T, and calculate for each one the sum of its length and the label on its S-vertex.

3. Choose the edge with the smallest sum. (If there are two or more with equally small sums, choose any of them at random.)

4. Label the T-vertex on that edge with the sum, and transfer it from T to S.

5. Repeat Steps 2 to 4 until the finish vertex has been transferred to S.

6. Find the shortest path by working backwards from the finish and choosing only those edges whose length is exactly equal to the difference of their vertex labels.

See how the algorithm works on the first example at the beginning of the chapter .

Initially, vertex O is in S, with label 0.

Edges joining {O} to {H, J, K, L, P} are OH (0+5=5), OK (0+14=14) and OL (0+17=17). Choose OH, label H as 5, and transfer H to set S.

Edges joining {O, H} to {J, K, L, P} are OK (14), OL (17), HJ (5+11=16) and HK (5+8=13). Choose HK, label K as 13, and transfer K to set S.

Edges joining {O, H, K} to {J, L, P} are OL (17), HJ (16), KJ (13+4=17), KP (13+20=33) and KL (13+3=16). There are two equal smallest sums, so choose (say) HJ, label J as 16, and transfer J to set S.

Edges joining {O, H, K, J} to {L, P} are OL (17), KL (16), KP (33) and JP (16+17=33). Choose KL, label L as 16, and transfer L to set S.

Edges joining {O, H, K, J, L} to {P} are KP (33), JP (33) and LP (16+16=32). Choose LP, label P as 32, and transfer P to set S.

Working back from P, the edges whose lengths are the difference between their vertex labels are PL, LK, KH and HO (HJ would be a forward, not a backward, move), so the shortest path is O - H - K - L - P, which is 32 miles long.

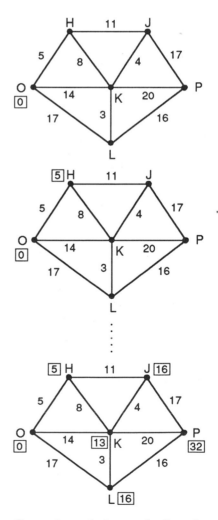

Some stages in the application of the algorithm

The second example can be dealt with similarly. In brief:

1. Label A as 0.

2. Choose AP and label P as 60.

3. Choose AQ and label Q as 90.

4. Choose PR and label R as 100.

5. Choose RU and label U as 190.

6. Choose UT and label T as 220.

7. Choose TS and label S as 280.

8. Choose SB and label B as 400.

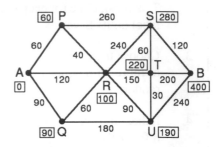

Scan back, choosing BS, ST, TU, UR, RP and PA to obtain the quickest route A - P - R - U - T - S - B, which takes 400 minutes.

Exercise 2A

1. Use the shortest path algorithm to find the shortest path from S to T in each of the diagrams below:

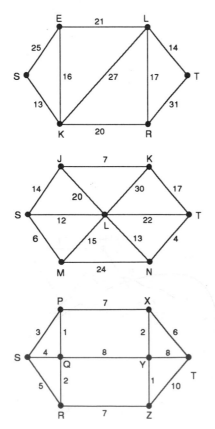

2. The table shows the cost in £ of direct journeys which are possible on public transport between towns A, P, Q, R, S and B. Find the cheapest route by which a traveller can get from A to B.

	A	P	Q	R	S	B
A	-	6	4	-	-	-
P	6	-	-	5	6	-
Q	4	-	-	3	7	-
R	-	5	3	-	-	8
S	-	6	7	-	-	5
B	-	-	-	8	5	-

3. The diagram represents the length in minutes of the rail journeys between various stations. Allowing 10 minutes for each change of trains, find the quickest route from M to N.

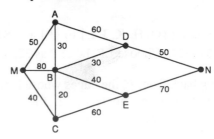

2.2 The minimum connector problem

A cable television company wants to provide a service to each of five towns, and for this purpose the towns must be linked (directly or indirectly) by cable. For reasons of economy, the company are anxious to find the layout that will minimise the length of cable needed.

The diagram shows in the form of a graph - not a map drawn to scale - the distances in miles between the towns. The layout of cable needs to form a connected graph joining up all the original vertices but, for economy, will not have any cycles. Hence we are looking for a subgraph of the one illustrated which is a tree and which uses all the original vertices: such a subgraph is called a **spanning tree**.

In addition, to find the minimum length of cable needed, from all the possible spanning trees we are looking for one with the total length of its edges as small as possible: this is sometimes known as a **minimum connector**.

The lower diagram shows one of the spanning trees of the graph. It is certainly a tree, and it joins all the vertices using some of the edges of the original graph. Its total length is 26 miles, however, and is not the shortest possible.

Five towns in Cleveland

Activity 4   Minimum-length spanning tree

Try to find another spanning tree whose total length is minimal.

An oil company has eight oil rigs producing oil from beneath the North Sea, and has to bring the oil through pipes to a terminal on shore. Oil can be piped from one rig to another, and rather than build a separate pipe from each rig to the terminal the company plans to build the pipes in such a way as to minimise their total length.

Activity 5

If the distances in km between the rigs A - H and the terminal T are as shown in the table on the next page, find this minimum total length and say how the connections should be made in order to minimise the total length.

	T	A	B	C	D	E	F	G	H
T	-	120	150	140	120	100	160	70	180
A	120	-	60	60	90	190	210	160	40
B	150	60	-	20	80	180	170	160	50
C	140	60	20	-	40	160	150	140	60
D	120	90	80	40	-	130	70	110	120
E	100	190	180	160	130	-	140	30	220
F	160	210	170	150	70	140	-	150	200
G	70	160	160	140	110	30	150	-	200
H	180	40	50	60	120	220	200	200	-

Like most examples in mathematics books, these problems are oversimplified in various ways. The presentation of the problems implies that junctions can occur only at the vertices already defined, and while this may just possibly be true in the second case (because of the difficulty and expense of maintaining a junction placed on the sea bed away from any rig) it seems unlikely in the first. On the other hand, the terms of the second problem ignore the fact that the costs of laying any pipe depend to some extent on the volume of oil that it is required to carry, and not solely on its length. Even so, these examples serve to illustrate the general principles involved in the solution of real problems.

The first problem above, with just five vertices, can be solved by a combination of lucky guesswork and common sense. The second problem is considerably harder - with nine vertices it is still just about possible to find the minimum connector by careful trial and error, but it is certainly not straightforward. As the number of vertices increases further, so the need for an algorithm grows.

Try to write down your own algorithm for solving problems of this type.

2.3   Kruskal's algorithm

There are two generally-known algorithms for solving the minimum connector problem; your algorithm may be essentially the same as one of these, or it may take a different approach altogether. Whatever the case, work carefully through the next two sections to see how Kruskal's and Prim's algorithms work, and compare the results they give with any that you may have obtained by other methods.

You will recall that the problem is to find a minimum-length spanning tree, and that a spanning tree is a subgraph including all the vertices but (because it is a tree) containing no cycles. **Kruskal's algorithm**, sometimes known as the 'greedy algorithm', makes use of these facts in a fairly obvious way.

The algorithm for n vertices is as follows:

1. Begin by choosing the shortest edge.

2. Choose the shortest edge remaining that does not complete a cycle with any of those already chosen. (If there are two or more possibilities, choose any one of them at random.)

3. Repeat Step 2 until you have chosen $n - 1$ edges altogether; the result is a minimum-length spanning tree.

Example

Look again at the first problem posed in Section 2.2, the diagram for which is repeated here for convenience.

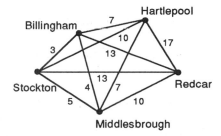

The shortest edge is BS = 3, so choose BS.

The shortest edge remaining is BM = 4, so choose BM.

The shortest edge remaining is SM = 5, but this completes a cycle and so is not allowed. Next shortest is BH or MH, both = 7; so choose randomly (say) BH.

The shortest edge remaining (apart from SM, which is already excluded) is HM, but this would complete a cycle; choose MR as the next shortest.

The four edges now chosen form a spanning tree of total length 24 miles, and this is the solution.

Example

The second problem, for which the distance table is repeated overleaf, can be solved using the same algorithm. You may find it helpful to draw a diagram and mark on the edges as they are chosen. A computer could not do this, of course.

How does a computer identify (and avoid) cycles in applying this algorithm?

	T	A	B	C	D	E	F	G	H
T	-	120	150	140	120	100	160	70	180
A	120	-	60	60	90	190	210	160	40
B	150	60	-	20	80	180	170	160	50
C	140	60	20	-	40	160	150	140	60
D	120	90	80	40	-	130	70	110	120
E	100	190	180	160	130	-	140	30	220
F	160	210	170	150	70	140	-	150	200
G	70	160	160	140	110	30	150	-	200
H	180	40	50	60	120	220	200	200	-

The shortest edge is BC (20), so choose BC.

The shortest remaining is EG (30), so choose EG.

The shortest remaining is AH (40) or CD (40), so choose (say) AH at random.

The shortest remaining is CD (40), so choose CD.

The shortest remaining is BH (50), so choose BH.

The shortest remaining are AB (60), AC (60) and CH (60), but any of these would complete a cycle; the next shortest is TG (70) or FD (70), so choose TG at random.

The shortest remaining (apart from those already excluded) is FD (70), so choose FD.

The shortest remaining is BD (80) but this completes a cycle, as does AD (90); the next shortest is DG (110), so choose DG.

There are nine vertices altogether, so the eight edges now chosen form a minimum-length spanning tree with total length 430 km.

Exercise 2B

1. The owner of a caravan site has caravans positioned as shown in the diagram, with distances in metres between them, and wants to lay on a water supply to each of them. Use Kruskal's algorithm to determine how the caravans should be connected so that the total length of pipe required is a minimum.

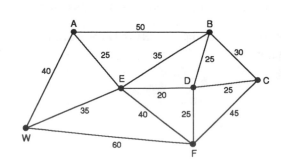

2. The warden of an outdoor studies centre wants to set up a public address system linking all the huts. The distances in metres between the huts are shown in the diagram opposite. How should the huts be linked to minimise the total distance?

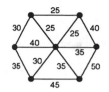

3. A complicated business document, currently written in English, is to be translated into each of the other European Community languages. Because it is harder to find translators for some languages than for others, some translations are more expensive than others; the costs in ECU are as shown in the table opposite.

Use Kruskal's algorithm to decide which translations should be made so as to obtain a version in each language at minimum total cost.

From / To	Dan	Dut	Eng	Fre	Ger	Gre	Ita	Por	Spa
Danish	-	90	100	120	60	160	120	140	120
Dutch	90	-	70	80	50	130	90	120	80
English	100	70	-	50	60	150	110	150	90
French	120	80	50	-	70	120	70	100	60
German	60	50	60	70	-	120	80	130	80
Greek	160	130	150	120	120	-	100	170	150
Italian	120	90	110	70	80	100	-	110	70
Portuguese	140	120	150	100	130	170	110	-	50
Spanish	120	80	90	60	80	150	70	50	-

2.4 Prim's algorithm

Although Kruskal's algorithm is effective and fairly simple, it does create the need to check for cycles at each stage. This is easy enough when calculations are being done 'by hand' from a graph, but (as you may have discovered) is less easy when working from a table and is quite difficult to build into a computer program.

An alternative algorithm which is marginally harder to set out, but which overcomes this difficulty, is **Prim's algorithm**. Unlike Kruskal's algorithm, which looks for short edges all over the graph, Prim's algorithm starts at one vertex and builds up the spanning tree gradually from there.

The algorithm is as follows:

1. Start with any vertex chosen at random, and consider this as a tree.

2. Look for the shortest edge which joins a vertex on the tree to a vertex not on the tree, and add this to the tree. (If there is more than one such edge, choose any one of them at random.)

3. Repeat Step 2 until all the vertices of the graph are on the tree; the tree is then a minimum-length spanning tree.

Example

Consider once again the cable television problem from Section 2.2.

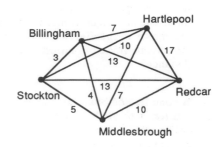

For no particular reason, choose Hartlepool as the starting point.

The shortest edge joining {H} to {B, S, M, R} is HB or HM, so choose either of these - HM, say - and add it to the tree.

The shortest edge joining {H, M} to {B, S, R} is MB, so add that to the tree.

The shortest edge joining {H, M, B} to {S, R} is BS, so add that to the tree.

The shortest edge joining {H, M, B, S} to {R} is MR, so add that to the tree.

All the vertices are now on the tree, so it is a spanning tree which (according to the algorithm) is of minimum length.

Example

Similarly, consider again the problem about the oil wells at distances as in the table:

	T	A	B	C	D	E	F	G	H
T	-	120	150	140	120	100	160	70	180
A	120	-	60	60	90	190	210	160	40
B	150	60	-	20	80	180	170	160	50
C	140	60	20	-	40	160	150	140	60
D	120	90	80	40	-	130	70	110	120
E	100	190	180	160	130	-	140	30	220
F	160	210	170	150	70	140	-	150	200
G	70	160	160	140	110	30	150	-	200
H	180	40	50	60	120	220	200	200	-

The obvious place to start is the terminal, so place T on the tree.

The shortest edge between {T} and {A, B, C, D, E, F, G, H} is TG, so add this to the tree.

The shortest edge between {T, G} and {A, B, C, D, E, F, H} is GE, so add this.

The shortest edge between {T, G, E} and {A, B, C, D, F, H} is GD.

The shortest edge between {T, G, E, D} and {A, B, C, F, H} is DC.

The shortest edge between {T, G, E, D, C} and {A, B, F, H} is CB.

The shortest edge between {T, G, E, D, C, B} and {A, F, H} is BH.

The shortest edge between {T, G, E, D, C, B, H} and {A, F} is HA.

The shortest edge between {T, G, E, D, C, B, H, A} and {F} is DF.

The tree now includes all the vertices, and thus is the minimum-length spanning tree required.

Exercise 2C

1. Use Prim's algorithm to solve the minimum connector problem for each of the graphs below.

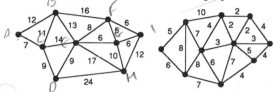

2. A group of friends wants to set up a message system so that any one of them can communicate with any of the others either directly or via others in the group. If their homes are situated as shown below, with distances in metres between them as marked, use Prim's algorithm to decide where they should make the links so that the total length of the system will be as small as possible.

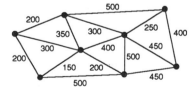

3. The chief greenkeeper of a nine-hole golf course plans to install an automatic sprinkler system using water from the mains at the clubhouse to water the greens. The distances in metres between the greens are as shown in the table below. Use Prim's algorithm to decide where the pipes should be installed to make their total length a minimum.

	CH	1	2	3	4	5	6	7	8
1	250								
2	600	350							
3	400	50	300						
4	100	200	600	300					
5	500	350	200	150	400				
6	800	550	200	500	750	350			
7	600	400	150	300	550	250	150		
8	350	200	350	100	350	50	400	250	
9	50	300	600	400	100	500	850	600	350

[If you want further examples for practice, repeat Exercise 2B using Prim's algorithm instead of Kruskal's algorithm.]

2.5 The travelling salesman problem

A problem not too dissimilar from the minimum connector problem concerns a travelling salesman who has to visit each of a number of towns before returning to his base. For obvious reasons the salesman wants to take the shortest available route between the towns, and the problem is simply to identify such a route.

This can be treated as a problem of finding a minimum-length Hamiltonian cycle, even though in practice the salesman has slightly more flexibility. A Hamiltonian cycle visits every vertex exactly once, you may recall, but there is nothing to stop the salesman passing through the same town twice - even along the same road twice - if that happens to provide the shortest route. This minor difference is rarely of any consequence, however, and for the purpose of this chapter it is ignored.

Activity 6

A salesman based in Harlow in Essex has to visit each of five other towns before returning to his base. If the distances in miles between the towns are as shown on the diagram opposite, in what order should he make his visits so that his total travelling distance is a minimum?

Find a solution by any method, and compare your answer with those of other students.

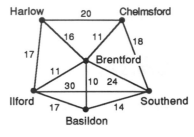

Southern Essex

The shortest route is actually a Hamiltonian cycle in this instance, and has a total length of 90 miles. It can be solved intuitively: there is a fairly obvious 'natural' route around the five outer towns, and all that remains is to decide when to detour via Brentford. For six towns arranged as these are, such a method is generally good enough.

Computer search

An alternative method, impractical for pencil-and-paper calculation but not unreasonable for a computer, is simply to find the total length of every possible route in turn. Given that the route must start and finish at Harlow, there are only 120 different orders in which to visit the other five towns (assuming than no town is visited more than once) and the necessary calculations can be completed in no more than a minute or two. Only half the routes really need to be checked, since the other half are the same routes in reverse, but if the time is available it is much simpler to check them all.

Take a moment or two to think about the procedure that might be used to take the computer through the 120 different routes.

The most obvious way is to keep the first town constant while the computer runs through the 24 possibilities for the other four (which it does by keeping the first of them constant while it goes through the 6 possibilities for the other three, etc.), doing this for each of the five possible first towns. A clever programmer could reduce such a scheme to just a few lines of code using recursive functions - that is, functions defined in terms of themselves!

A less obvious alternative approach is by analogy with bellringing. Traditional English church bellringers do not play tunes, but instead ring 'changes' by ringing the bells in every possible order - essentially the same as the problem here. Many different 'systems' have been devised for ringing changes, with names such as 'Plain Bob', 'Steadman' and 'Grandsire', and any of these can be converted without too much difficulty to a computer program.

Now it is clear that if a computer can be used to check every possible route and find the shortest, that route will be the solution to the salesman's problem. That might appear to be an end to the matter, but in fact it is not. For six towns there are only 120 different routes to try, and that is quite manageable even on a desktop computer, but as the number of towns increases the number of possible routes increases factorially. Thus for ten towns there are 362880 possible routes, for fifteen there are nearly a hundred billion, and for twenty there are more than 10^{17}. Even the fastest computers would take many years, if not many centuries, to check all the possible routes around twenty towns, making the full search method of little use in practice.

Further examples

Example

A business executive based in London has to visit Paris, Brussels and Frankfurt before returning to London. If the journey times in hours are as shown in the diagram opposite, work out the total length of every possible Hamiltonian cycle and thus find the route that takes the shortest time.

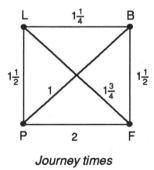

Journey times

There are three cities apart from London, and so $3 \times 2 \times 1 = 6$ possible orders. The six journeys with their total lengths are as follows:

$$L - P - B - F - L = 5\tfrac{3}{4}\,h \qquad L - F - B - P - L = 5\tfrac{3}{4}\,h$$

$$L - P - F - B - L = 6\tfrac{1}{4}\,h \qquad L - B - F - P - L = 6\tfrac{1}{4}\,h$$

$$L - B - P - F - L = 6\,h \qquad L - F - P - B - L = 6\,h$$

It is clear from this that the shortest route is

London - Paris - Brussels - Frankfurt - London

or the same in reverse.

Example

A visitor to the County Show wants to start from the main gate, visit each of eight exhibitions, and return to the main gate by the shortest possible route. The distances in metres between the exhibitions are given in the table below. What route should the visitor take?

Gate	A	B	C	D	E	F	G	H
Gate	-	200	350	400	500	350	150	200	350
A	200	-	200	300	400	450	300	250	200
B	350	200	-	100	250	450	500	300	100
C	400	300	100	-	150	350	500	300	100
D	500	400	250	150	-	250	450	300	200
E	350	450	450	350	250	-	250	200	350
F	150	300	500	500	450	250	-	200	400
G	200	250	300	300	300	200	200	-	200
H	350	200	100	100	200	350	400	200	-

Without a graph this problem is very difficult, but it is possible to make some sort of attempt at a solution.

Activity 7

Try to find a minimum-length route and compare your answer with those of other students.

The 'best' solution has a total length of 1550 m. This is the length of the route Gate - A - H - B - C - D - E - G - F - Gate, but there are other routes of the same minimum length. If you found any of these routes for yourself, you should feel quite pleased.

You might expect at this point to be given a standard algorithm for the solution of the travelling salesman problem. Unfortunately, no workable general algorithm has yet been discovered - the exhaustive search method is reliable but (for large vertex sets) takes too long to be of any practical use. Intuitive 'common sense' methods can often lead to the best (or nearly best) solution in particular cases, but the general problem has yet to be solved.

* Upper and lower bounds

Although there is no general algorithm for the solution of the travelling salesman problem, it is possible to find upper and lower bounds for the minimum distance required. This can sometimes be very useful, because if you know that the shortest route is between (say) 47 miles and 55 miles long, and you can find a route of length 47 miles, you know that your answer is actually a solution. Alternatively, from a business point of view, if the best route you can find by trial-and-error is 48 miles long, you might well decide that the expense of looking for a shorter route was just not worthwhile.

Finding an upper bound is easy: simply work out the length of any Hamiltonian cycle. Since this cycle gives a possible solution, the best solution must be no longer than this length. If the graph is not too different from an ordinary map drawn to scale, it is usually possible by a sensible choice of route to find an upper bound quite close to the minimum length.

Finding a good lower bound is a little trickier, but not impossible. Suppose that in a graph of 26 vertices, A - Z, you had a minimum length Hamiltonian cycle, AB, BC, CD, ..., YZ, ZA. If you remove any one of its vertices, Z say, and look at the remaining graph on vertices A - Y then

minimum length of a Hamiltonian cycle

= length of AB, BC, CD, ..., XY, YZ, ZA

= (length of AB, BC, CD, ..., XY) + (length of YZ) + (length of ZA)

$$\geq \left(\begin{array}{c}\text{minimum length of a spanning tree} \\ \text{of the graph on vertices A - Y}\end{array}\right) + \left(\begin{array}{c}\text{lengths of the two} \\ \text{shortest edges from Z}\end{array}\right)$$

So a lower bound for the minimum-length Hamiltonian cycle is given by the minimum length of a spanning tree of the original graph without one of its vertices, added to the lengths of the two shortest edges from the remaining vertex.

Look at the County Show problem on pages 29/30 and see how these two methods work. An upper bound is easy to find: the cycle

Gate -A - B -C- D - E - F - G - H - Gate

has total length 1900 m, so the optimum solution must be no more than this. Simply by trial and error, you may be able to find a shorter route giving a better upper bound.

Now let us consider the problem of finding a good lower bound by the method just given. Suppose the vertex E is removed. Using Prim's algorithm as in Section 2.4, the minimum-length spanning tree for the remaining vertices has length 1100 m. The two shortest edges from E are 200 and 250, and adding these to the spanning tree for the rest gives 1550 m. The optimum solution must have at least this length, but may in fact be longer - there is no certainty that a route as short as this exists. The best route around the Show thus has a length between 1550 m and 1900 m.

This technique gives no indication as to how such a best route can be found, of course, although a minimum-length spanning tree may be a useful starter. But if by clever guesswork you can find a route equal in length to the lower bound, you can be certain that it is in fact a minimum-length route. The route Gate - A - H - B - C - D - E - G - F - Gate is an example: it has length 1550 m, equal to the lower bound, and so is certainly a minimum.

Exercise 2D

The following problems may be solved by systematic search by hand or computer, by finding upper and lower bounds, by trial and error, or in any other way.

1. Find a minimum-length Hamiltonian cycle on the graph shown in the diagram below.

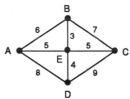

2. The Director of the Scottish Tourist Board, based in Edinburgh, plans a tour of inspection around each of her District Offices, finishing back at her own base. The distances in miles between the offices are shown in the diagram; find a suitable route of minimum length.

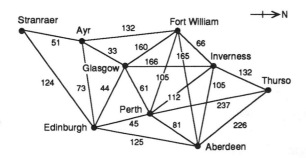

3. A milling machine can produce four different types of component as long as its settings are changed for each type. The times in minutes required to change settings are shown in the table.

From / To	A	B	C	D	Off
A	-	5	7	4	8
B	5	-	6	7	6
C	7	6	-	5	9
D	4	7	5	-	7
Off	8	6	9	7	-

On a particular day, some of each component have to be produced. If the machine must start and finish at 'Off', find the order in which the components should be made so that the time wasted in changing settings is as low as possible.

2.6 The Chinese postman problem

The Chinese postman problem takes its name not from the postman's nationality, but from the fact that it was first seriously studied by the Chinese mathematician *Mei-ko Kwan* in the 1960s. It concerns a postman who has to deliver mail to houses along each of the streets in a particular housing estate, and wants to minimise the distance he has to walk.

How does this differ from the problem of the travelling salesman?

The travelling salesman wants to visit each town - each vertex, to use the language of graph theory - so the solution is a Hamiltonian cycle. The postman, on the other hand, wants to travel along each road - each edge of the graph - so his problem can best be solved by an Eulerian trail if such a thing exists. Most graphs are not Eulerian, however, and this is what makes the problem interesting.

The diagram shows a housing estate with the length in metres of each road. The postman's round must begin and end at A, and must take him along each section of road at least once.

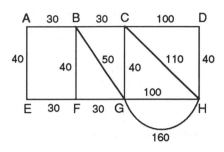

Activity 8

Try to find a route of minimum total length.

Once again, trial and error will often lead to a satisfactory solution where the number of vertices is small, but for larger graphs an algorithm is desirable. A complete algorithm for the solution of the Chinese postman problem does exist, but it is too complicated to set out in full here. What follows is a much simpler partial algorithm that will work reasonably well in most cases.

Systematic solution

The algorithm combines the idea of an Eulerian trail with that of a shortest path. You will recall that an Eulerian graph can be identified by the fact that all its vertices have even degree, and this is at the heart of the systematic solution. The method is as follows:

Find the degree of each vertex of the graph.

1. If all the vertices are of even degree, the graph is Eulerian and any Eulerian trail is an acceptable shortest route.

2. If just two vertices are of odd degree, use the algorithm from Section 2.1 to find the shortest path between them; the postman must walk these edges twice and each of the others once.

3. If more than two vertices are of odd degree - this is where the partial algorithm fails - use common sense to look for the shortest combination of paths between pairs of them. These are the edges that the postman must walk twice.

Example

Consider again the problem above (regarding the layout of roads as a graph with vertices A - H). The only two odd vertices are F and G, and the shortest path between them is obviously the direct edge FG. The postman must therefore walk this section of road twice and all the rest once: a possible route would be

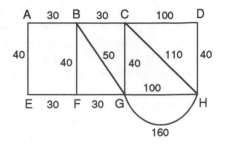

A - B - C - D - H - G - H - C - G - B - F - G - F - E - A.

The total length of any such shortest route is the sum of all the edge lengths plus the repeated edge, which is 830 m.

Example

A roadsweeper has to cover the road system shown in the diagram opposite, going along every road at least once but travelling no further than necessary. What route should it take?

There are four odd vertices, E, G, H and J. By common sense, the shortest combined path comes from joining H to G directly, and J to E via F, so the repeated edges must be HG, FJ and FE with a combined length of 120 m. The edges of the graph have total length 800 m, so the sweeper's best route is 920 m long. One such route is

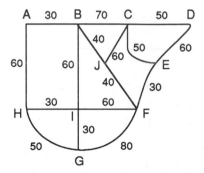

A - B - C - D - E - C - J - F - E - F - I - B - J - F - G - I - H - G - H - A

but there are many other equally valid routes.

Exercise 2E

1. Find a solution to the Chinese postman problem on each of the graphs below.

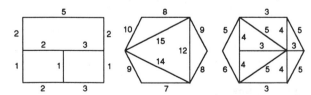

2. A church member has to deliver notices to the houses along each of the roads shown in the diagram below.

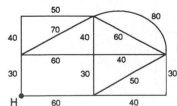

If her own house is at H, what route should she follow in order to make her total walking distance as short as possible?

3. After a night of heavy snow, the County Council sends out its snow plough to clear the main roads shown (with their lengths in km) in the diagram below.

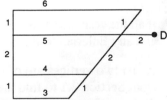

The plough must drive at least once along each of the roads to clear it, but should obviously take the shortest route starting and finishing at the depot D. Which way should it go?

2.7 Local applications

The previous sections of this chapter have covered four classes of problem: the shortest path problem, the minimum connector problem, the travelling salesman problem and the Chinese postman problem. They have some similarities, but each class of problem requires a slightly different method for its solution, and it is important to recognise the different problems when they occur.

One way of acquiring the ability to distinguish the four problems from one another is to consider a selection of real life problems and try to classify them - even try to solve them, if they are not too complicated. Time set aside for this purpose will not be wasted.

The problems you find will depend on your own local environment, but might include

- finding the quickest way of getting from one point in the city to another, on foot, or on a cycle, or by car, or by bus;

- finding the shortest route around your school or college if you have a message to deliver to every classroom;

- finding the shortest route that takes you along every line of your nearest Underground or Metro system, and trying it out in practice;

- planning a hike or expedition visiting each of six pre-chosen churches, or hilltops, or other schools, or pubs;

- whatever catches your imagination.

2.8 Miscellaneous Exercises

1. Use Kruskal's or Prim's algorithm to find a minimum-length spanning tree on the graph below.

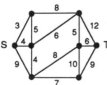

2. Use the algorithm given to find the shortest path from S to T on the graph above.

3. Solve the Chinese postman problem for the graph shown in the diagram above.

4. Find a minimum-length Hamiltonian cycle on the graph above.

5. A railway track inspector wishes to inspect all the tracks shown in the diagram below, starting and finishing at the base B (distances shown in km).

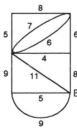

Explain why this cannot be done without going over some sections of track more than once, and find the shortest route the inspector can take.

6. A building society with offices throughout Avon wants to link its branches in a private computer network.

   The distances in miles between the branches are as shown in the table opposite.

   Find a way of connecting them so that the total length of cable required is a minimum.

	BA	BR	CS	CL	KE	KI	LU	PA	PO	RA	WE
Bath		13	12	23	7	9	16	16	22	8	29
Bristol			11	12	6	5	8	7	9	14	21
Chipping Sodbury				23	11	7	19	8	19	20	32
Clevedon					17	17	8	16	5	22	8
Keynsham						4	10	11	15	9	23
Kingswood							13	7	14	13	26
Luisgate								15	8	14	13
Patchway									11	20	24
Portishead										21	13
Radstock											24
Weston Super Mare											

7. A sales rep based in Bristol has to visit shops in each of seven other towns before returning to her base. The distances between the towns are as shown in the diagram below.

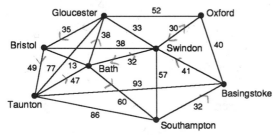

   Find a suitable route of minimum length.

8. The diagram below shows the various possible stages of an air journey, each marked with its cost in dollars.

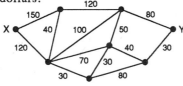

   Use a suitable algorithm to find the least expensive route from X to Y, and state its cost.

9. The groundsman of a tennis club has to mark out the court with white lines, with distances in feet as shown in the diagram. Because the painting machine is faulty, it cannot be turned off and so must go only along the lines to be marked. How far must the groundsman walk, given that he need not finish at the same place he started?

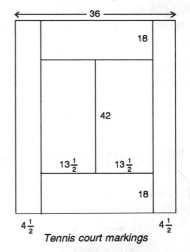

   Tennis court markings

10. An American tourist arrives at Heathrow and wants to visit London, Oxford, Cambridge, Stratford-on-Avon, Grasmere, Edinburgh, Bath, and her ancestors' home in Norwich before returning to Heathrow to catch a flight back to the USA.

   The travel times in hours between places are as shown in the table opposite.

   If she wants to spend 6 hours in each place, can she complete such a journey in the 75 waking hours she has available before her flight leaves?

	L	O	C	S	G	E	B	N	H
London	-	$1\frac{1}{2}$	$1\frac{1}{2}$	3	5	$4\frac{1}{2}$	2	3	1
Oxford	$1\frac{1}{2}$	-	3	$1\frac{1}{2}$	6	6	2	$4\frac{1}{2}$	$1\frac{1}{2}$
Cambridge	$1\frac{1}{2}$	3	-	4	6	$5\frac{1}{2}$	$3\frac{1}{2}$	2	$2\frac{1}{2}$
Stratford	3	$1\frac{1}{2}$	4	-	5	$6\frac{1}{2}$	$3\frac{1}{2}$	$5\frac{1}{2}$	3
Grasmere	5	6	6	5	-	4	7	8	6
Edinburgh	$4\frac{1}{2}$	6	$5\frac{1}{2}$	$6\frac{1}{2}$	4	-	$6\frac{1}{2}$	7	$5\frac{1}{2}$
Bath	2	2	$3\frac{1}{2}$	$3\frac{1}{2}$	7	$6\frac{1}{2}$	-	5	2
Norwich	3	$4\frac{1}{2}$	2	$5\frac{1}{2}$	8	7	5	-	4
Heathrow	1	$1\frac{1}{2}$	$2\frac{1}{2}$	3	6	$5\frac{1}{2}$	2	4	-

11. The following table gives the distances between five towns, A, B, C, D and E. Use the greedy algorithm to construct a minimum connector joining these towns.

	A	B	C	D	E
A	-	9	3	7	6
B	9	-	8	9	8
C	3	8	-	7	7
D	7	9	7	-	4
E	6	8	7	4	-

12. Use the shortest path algorithm to find the shortest path from S to T in the following network.

(Your answer should show clearly how the algorithm is being applied, and what the vertex-labels are at each stage.)

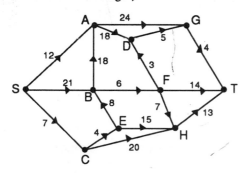

13. The network in the diagram below indicates the main road system between 10 towns and cities in the North of England.

   (a) A computer company wishes to install a computer network between these places, using cables laid alongside the roads and designed so that all places are connected (either directly or indirectly) to the main computer located at Manchester. Find the network which uses a minimum total length of cable.

   (b) A highways maintenance team, based at Manchester, wishes to inspect all the roads shown in the network at least once. Design the route for them, which starts and finishes at Manchester, and has the smallest total distance.

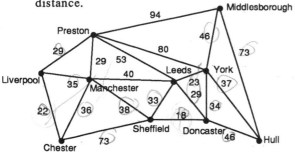

14. Use the shortest-path algorithm to find the shortest path between S and T for the following network.

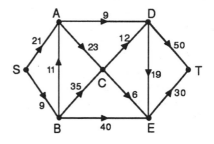

The numbers represent the lengths of each arc and you can only move in the directions indicated by the arrows.

15. The following table gives the distance (in km) between six water pumping stations.

	A	B	C	D	E	F
A	-	74	64	98	83	144
B	74	-	81	145	144	161
C	64	81	-	50	100	80
D	98	145	50	-	83	63
E	83	144	100	83	-	151
F	144	161	80	63	151	-

   (a) Use the greedy algorithm to construct a minimum connector between the stations.

   (b) Explain how the greedy algorithm can be used in general to provide a lower bound for the solution of the travelling salesman problem, and obtain such a lower bound for the six pumping stations.

   (c) Find an upper bound for this problem.

16. The graph below represents a railway network connecting towns A, B, C, D, E, F, and G, with distances between towns shown in kilometres.

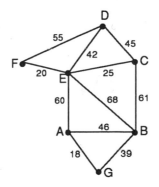

An inspection train is required to begin at B and survey all of the track and return to B. The train needs to travel only in one direction along any section of track in order to carry out the survey of that section.

(a) Explain why it is impossible to complete the survey by travelling along each section of track once and only once.

(b) Find a route of minimum length for the train, stating which section(s) of track must be repeated, and determine the total route length.

(c) Is the solution found in (b) unique? Give an explanation for your answer. (AEB)

17. The network below illustrates the main junctions on a city rail sytem.

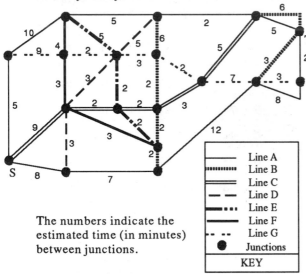

The numbers indicate the estimated time (in minutes) between junctions.

(a) Use an appropriate shortest path algorithm to find the path of minimum total time between S and T, assuming that there is no waiting when changing trains at junctions.

(b) Assuming that an average time for changing and waiting at a junction is 5 minutes, is the solution determined in (a) still valid? If not, what is now the optimal solution?

(Note that, at a junction, you only change trains if you are changing from one line to a different one.) (AEB)

18. The network below illustrates the time (in minutes) of rail journeys between the given cities.

(a) Assuming that any necessary changes of train do not add any time to the journey, find the quickest train route from Manchester to Cambridge.

(b) Assuming that at each city shown a change of train is always necessary, and that each change adds 20 minutes to the journey time, find the quickest train route from Manchester to Cambridge.

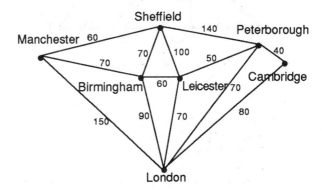

(AEB)

3 ENUMERATION

Objectives

After studying this chapter you should

* understand the basic principles of enumeration;

* know and be able to use the $\begin{pmatrix} n \\ r \end{pmatrix}$ notation;

* be able to apply this notation to the solution of various problems.

3.0   Introduction

Enumeration is really just a fancy name for counting, and the aim of this chapter is to teach you how to count! But that's not quite as silly as it may seem, since enumeration is concerned particularly with counting the number of ways in which something can be done - often, the number of ways in which a particular choice can be made.

There is an important linguistic convention here. When a mathematician asks (for example) how many ways there are of choosing three students from a class of 21, she doesn't want a list like

- choose the three oldest

- choose the first three on the register

- use random selection

- let them fight for it

and so on. This might be something to discuss, but is not enumeration.

The question "how many ways?" in this area of mathematics is understood to refer not to the mechanism of choosing but to the number of different results. Thus the choices could be

- Mary, John and Jasmine, or

- Mary, Lloyd and Martin, or

- Trevor, Mary and Desmond,

or any of 1327 other possibilities. You will see shortly how this figure can be calculated.

One final point before you begin to study this topic. We have chosen to use the title **Enumeration** to describe it, but this name is not universal. Some textbooks use the label **combinatorics** to describe enumeration (or often to describe enumeration and graph theory taken together), while many others use the older title **permutations** and **combinations**. If you are using other books alongside this one (as you should, from time to time), you need to be familiar with this variation.

3.1 The multiplicative principle

A restaurant offers a **Special Business Lunch** menu as shown opposite for the price of £3.95.

How many different meals could be served, each chosen from this menu?

You might start by trying to list all the meals:

Soup, Curry, Fruit salad

Soup, Liver, Cheese

Juice, Curry, Ice cream

Soup, Chow Mein, Cheese ....

That is clearly not going to be very reliable, because if the meals are listed in this almost random order it is very difficult to know whether or not you have included every possibility exactly once.

So a better list might be in some logical order:

Juice, Curry, Fruit salad

Juice, Curry, Ice cream

Juice, Curry, Cheese

Juice, Chow mein, Fruit salad

....

Juice, Liver, Cheese

Soup, Curry, Fruit salad

....

Soup, Liver, Cheese

What meals will fill the gaps? How many meals are there altogether?

What you will find if you write out the full list - you may be able to see it already in your imagination - is that because there are three 'desserts' there will be three meals for each combination of starter and main course; and because there are three main courses there will be three such combinations for each starter; and there are two starters.

So the total number of different meals is

2 (starters) × 3 (main courses) × 3 (desserts) = 18.

Special Business Lunch

Fruit Juice
or
Soup of the Day

———

Chicken Curry with Rice
or
Pork Chow Mein
or
Liver with Onions and
Mashed Potatoes

———

Fruit Salad and Cream
or
Ice Cream
or
Cheese and Biscuits

£3.95

This is an example of what is often called the **multiplicative principle.** Where several independent choices have to be made one after another, and there are r options for the first choice, s options for the second choice, t options for the third choice, and so on, the total number of possibilities is

$$r \times s \times t \times \ ...$$

Example

Sandra has six different sweaters and five different pairs of trousers. How many different combinations of these can she wear?

Solution

Assuming Sandra chooses sweater and trousers independently, she has $6 \times 5 = 30$ possibilities. In reality there are probably some particular combinations (the pink sweater with the orange trousers, for example) that she wouldn't be seen dead in! But even that can be taken into account as long as the limitations are spelled out in advance.

Example

In a tennis club there are 16 men (three of whom are County players) and 23 women (including six County players). In how many ways can the club select a mixed doubles team for a competition, if it must not include more than one County player?

Solution

First of all, ignore the restriction. The club must choose one of 16 men, then one of 23 women, so there are $16 \times 23 = 368$ possible pairs. But some of these pairs contain two County players, which is illegal. In fact there are three male and six female County players, so there are $3 \times 6 = 18$ pairs in which both players are County players.

So there are $368 - 18 = 350$ legal mixed doubles teams the club can choose.

Exercise 3A

1. A group of friends go into a cafe together. Each of them orders either tea or coffee, with or without milk, with no sugar, one sugar or two sugars. If no two of them have exactly the same, how many friends could there be?

2. How many different six-figure telephone numbers are possible, if the first figure cannot be 0, 1 or 9?

3. An advertisement for a computer printer claims that it offers more than 100 different fonts. When you read the small print, you find that each combination of style, emphasis and size has been counted separately. If each letter can have any of four emphases (normal, italic, bold and italic bold) and any of three sizes (6 cpi, 12 cpi and 24 cpi), how many styles must there be?

4. An examination paper is divided into three sections. Each section contains six questions, three of which are starred. The rubric says "Answer only one question from each section. Answer at least one starred question and at least one unstarred question." How many different choices are possible?

5. The 37 members of a club have to elect from among themselves a chairman, a secretary and a treasurer. Assuming no one can be elected to more than one of these positions, how many different results are possible?

3.2 Arrangements

Holy Trinity Church in Seaton Carew has four bells, which are rung before the main service every Sunday morning. Traditionally, English bellringers don't actually play tunes on the bells; instead, they 'play maths' by trying to ring the bells in every possible order.

How many possible orders are there for four bells, each rung once? This problem can be solved in a similar way to the last question of the exercise above. There are four choices for the bell that comes first, then three choices for the bell that comes second, then two choices for the bell that comes third, then one choice (!) for the bell that comes fourth. So by the multiplicative principle, there are $4 \times 3 \times 2 \times 1 = 24$ possible orders altogether.

Activity 1

Make a list of the 24 possible orders.

Then try to arrange these orders one after another so that no bell changes more than one place forward or back between one order and the next. That is,

$$1\ 3\ 4\ 2$$

$$1\ 4\ 3\ 2$$

is all right, because bells 3 and 4 each change just one place, while bells 1 and 2 do not move at all.

$$1\ 3\ 4\ 2$$

$$1\ 4\ 2\ 3$$

is not allowed, because bell 3 changes two places back.

The same basic method can be used to deal with other arrangements. If there are ten boats in a race, for example, the number of different orders in which they can finish (assuming dead heats are not allowed) is

$$10 \times 9 \times 8 \times 7 \times 6 \times 5 \times 4 \times 3 \times 2 \times 1 = 3\ 628\ 800.$$

Writing the multiplication out in full like this every time can obviously be very boring, and it is customary to adopt a mathematical shorthand:

N! (spoken as 'N factorial')

$$= N \times (N-1) \times (N-2) \times (N-3) \times \ldots \times 2 \times 1.$$

The table below shows the value of N! for values of N between 1 and 20.

1! =	1	11! =	39 916 800
2! =	2	12! =	479 001 600
3! =	6	13! =	6 227 020 800
4! =	24	14! =	87 178 291 200
5! =	120	15! =	1 307 674 368 000
6! =	720	16! =	20 922 789 888 000
7! =	5 040	17! =	355 687 428 096 000
8! =	40 320	18! =	6 402 373 705 728 000
9! =	362 880	19! =	121 645 100 408 832 000
10! =	3 628 800	20! =	2 432 902 008 176 640 000

Does 0! have a meaning? If not, why does your calculator give it a value?

Activity 2

Which grows faster as x increases: 10^x, or $x!$?

Don't jump to conclusions from just a few small-number values - see what happens when x gets quite large.

Activity 3

Calculating $x!$ for large values of x is quite time-consuming: even a pocket calculator gives up at 70! in most cases. But there is a formula called Stirling's formula which gives an approximate value of $x!$ for large x:

$$x! \approx \sqrt{(2\pi x)} \times \left(\frac{x}{e}\right)^x$$

where e ($\approx$ 2.718 ...) is the base of natural logarithms. Use the formula to calculate approximations to 20!, 50! and 100!, and compare them with the most accurate answers you can find elsewhere.

Using this factorial notation, it is easy to state a general rule for the number of possible arrangements of a finite set:

N different objects can be arranged in order in N! ways

Example

If six people have to share the six seats in a railway compartment, in how many ways can they sit?

Solution

Assuming no one sits on anyone else's lap, this is just a problem of arranging six different objects. This can be done in

$$6! = 6 \times 5 \times 4 \times 3 \times 2 \times 1 = 720 \text{ ways.}$$

Example

How many nine-letter words (which needn't actually make sense) can be made from the letters of SPREADING?

Solution

Nine different objects can be arranged in 9! = 362 880 orders.

Example

How many of these 'words' have all three vowels together?

Solution

Imagine the three vowels locked together as a single unit, called V, say. Then the problem is to arrange SPRDNGV, and this can be done in 7! ways. But the three vowels can be arranged in 3! ways among themselves, so altogether there are $7! \times 3! = 30\ 240$ possible arrangements.

If the objects are not all different, of course, the problem becomes more complicated.

Suppose the last example had asked for words formed from the letters of TOTTERING - what would the answer have been then?

There are still nine letters, but they are not all different. There are three identical Ts, and there will be fewer possibilities for rearrangement.

Suppose the Ts are made different by giving them labels - call them T_1, T_2 and T_3, say.

Now there are nine different letters once again, which can be formed into 362 880 words just as in the previous case. But look at some of these:

$T_1OT_2T_3ERING$	$T_1OT_3T_2ERING$	$T_2OT_1T_3ERING$
$T_2OT_3T_1ERING$	$T_3OT_1T_2ERING$	$T_3OT_2T_1ERING$

There are six apparent words here, each of them giving the same word TOTTERING once the labels are removed, so the arrangement TOTTERING has been counted six times among the 362 880. And this will be true for any other arrangement too, because for every arrangement the three Ts can be rearranged among themselves in $3! = 6$ ways.

So if every arrangement has been counted six times, the real number of words that can be formed from TOTTERING is

$$9! \div 3! = 362\ 880 \div 6 = 60\ 480$$

This idea can be extended to cases where there is more than one repetition, leading to the general rule:

If there are N objects, of which R are the same in one way, S the same in another way, and so on, the number of different arrangements is

$$\frac{N!}{R! \times S! \times \ldots}.$$

Example

In how many ways can the letters of MISSISSIPPI be arranged?

Solution

There are 11 letters here, including 4 Is, 4 Ss and 2 Ps. The total number of arrangements is therefore

$$11! \div (4! \times 4! \times 2!) = 39\ 916\ 800 \div 1152 = 34\ 650$$

Example

How many ways are there of getting from S to F in the diagram, by moving 'north' or 'east' only?

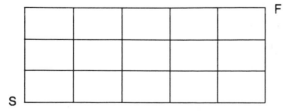

Solution

The journey from S to F involves five eastward moves and three northward moves, and so is equivalent to an arrangement of EEEEENNN. The number of such arrangements is

$$8! \div (5! \times 3!) = 56, \text{ so there are 56 possible routes.}$$

Exercise 3B

(Answers greater than 10^6 may be given correct to three significant figures.)

1. Eight chess players take part in a league. In how many possible orders could they finish?

2. If there are eighteen cricket teams in the County Championship, in how many possible orders could they finish the season?

3. In how many different ways can the thirteen cards in a hand be arranged?

4. Three red discs, two yellow discs, one blue disc and one green disc are to be put on a spike. In how many ways can this be done?

5. On a shelf in the Maths stockroom there are six copies of 'Pure Maths', four copies of 'Discrete Maths', two copies of 'Statistics', and one each of 'Mechanics', 'Further Pure Maths', and 'Further Statistics'. How many different arrangements of these books are possible?

6. How many words of eight letters can be formed from HOPELESS?

7. How many different five-digit numbers can be made from the digits 1, 2, 2, 3, 4? How many of these numbers are even?

8. In how many ways can the three medals be awarded in a race in which eight athletes take part?

9. Trevor, June, Melissa and Barry go to the cinema and sit in a row of four seats. How many possible seating arrangements are there if June and Barry refuse to sit next to one another?

*10. In how many ways can the letters of MATHEMATICS be arranged if no two vowels are adjacent?

Activity 4

When King Arthur and his twelve knights go in to dinner, in how many different ways can they sit around the Round Table?

How many different necklaces can be made by threading ten differently-coloured beads

  (i)   onto a chain with a large clasp, or

  (ii) onto a piece of thread tied into a circle with a very small knot?

What assumptions do you make in answering each of these questions? Discuss your assumptions and your answers with other students.

3.3 Making choices

A Sixth Form 'General Studies' programme offers thirty different courses, and each student is expected to choose three of them. How many different choices are possible?

A student has 30 possible first choices, 29 second choices, and 28 third choices, giving 24 360 possibilities altogether. But the assumption here is that the order of choosing is unimportant, because the choice 'Law, Pottery, Hockey' gives the same three courses as 'Hockey, Pottery, Law'. Each combination of three courses has thus been counted six times $(3! = 6)$, and the true number of combinations is $24\ 360 \div 6 = 4060$.

More generally, if there are n different items from which you have to choose r, where the order of choosing does not matter, the number of possible combinations is

$$n \times (n-1) \times (n-2) \times \ ... \ \times (n-r+1) \div r!$$

Example

In how many ways can you choose two cakes from a plate of eight?

Solution

$$8 \times 7 \div 2 = 28 \text{ ways.}$$

Example

In a particular company, employees can choose any four weeks (not necessarily consecutive) for their annual holiday. Last year, no two employees chose exactly the same four weeks; how many employees could there be?

Solution

If 4 weeks are chosen from 52, there are

$52 \times 51 \times 50 \times 49 \div 4! = 270 \ 725$ different combinations possible; there could be this many employees.

Activity 5

Working with other students (so that each of you need only calculate a few values), complete the following table to show the number of possible combinations of r objects chosen from n.

	$n = 2$	3	4	5	6
$r = 1$	?	?	4	?	?
2	1	?	?	10	?
3	—	1	?	?	20
4	—	—	1	?	?
5	—	—	—	1	?

Where have you seen the same numbers before, perhaps in a different orientation?

The number of ways of choosing r objects from n, where all the objects are different and the order of choosing is unimportant, is denoted by $\binom{n}{r}$ (spoken as "n choose r"), or in some older books by $^{n}C_{r}$ or $_{n}C_{r}$. Then

$$\binom{n}{r} = n \times (n-1) \times \; ... \; \times (n-r+1) \div r!$$

This is the same $\binom{n}{r}$ you may already have met in connection with the binomial theorem (in pure mathematics) or the binomial distribution (in statistics); you may like to discuss with your teacher the reasons for the similarity. If you calculator has a button labelled $^{n}C_{r}$, find out how to use it and check the answers from the examples above and below. Note also that you can write

$$\binom{n}{r} = \frac{n!}{(n-r)! \, r!}$$

Example

An environment group with 26 members has to choose three members to lobby their MP. How many possible delegations are there?

Solution

$$\binom{26}{3} = 2600 \text{ delegations.}$$

Example

In how many ways can ten basketball players be split into two teams of five?

Solution

$$\binom{10}{5} \div 2 = 126$$

The number of ways of choosing a single team is divided by 2 because choosing ABCDE gives the same two teams as choosing FGHIJ.

Example

In how many ways can a committee of four be chosen from eight men and nine women, if the committee must include at least one member of each sex?

Solution

There are 17 members, so $\binom{17}{4}$ possible committees altogether.

But $\binom{8}{4}$ of these committees are all male, and $\binom{9}{4}$ are all female, and neither of these is allowed. The number of valid choices is therefore $\binom{17}{4}-\binom{8}{4}-\binom{9}{4}=2380-70-126=2184$.

Exercise 3C

1. Evaluate the following:

   (a) $\binom{10}{3}$

   (b) $\binom{10}{6}$

   (c) $\binom{18}{4}$

   (d) $\binom{n}{n}$

   (e) $\binom{n}{0}$

2. In how many ways can Colleen choose three library books from a shelf of 22?

3. A cafeteria serves sausages, bacon, eggs, mushrooms, tomatoes, beans, hash browns and fried bread at breakfast time, and offers any five different items for £1.99. How many different breakfasts can be made up?

4. If there are 46 universities offering the course I want, in how many ways can I choose eight of them for my UCAS form?

5. To complete his pools coupon, Harry has to predict which 10 out of 58 football matches will end as 1-1 draws. In how many ways can he choose these ten matches?

6. In how many ways can a squad of eight be chosen from six sergeants and twelve other ranks, if the squad must include at least one sergeant?

7. From a squad of 16 players, a football team of 11 players must be chosen. In how many ways can this be done if only two of the squad can keep goal?

8. In how many ways can 16 athletes be divided into two equal teams for a tug of war?

9. Every week last term, Lucas was late to school on two of the five days, but never on the same two days. What is the longest the term could have been?

10. There are 77 applications for the 72 places on a coach trip. In how many ways can the 72 lucky applicants be chosen?

3.4 Arrangements

Activity 6

(a) Verify by several examples that for $n \geq r \geq 0$, $\binom{n}{r}=\binom{n}{n-r}$.

By considering different choosing procedures, explain why this should be true.

(b) Verify by several examples that for $n \geq r \geq 1$,

$\binom{n}{r}=\binom{n-1}{r}+\binom{n-1}{r-1}$. By considering different choosing procedures, explain why this result is true.

As with arrangements, the idea of choice can be extended to cases where the objects are not all different, though the extension is not particularly simple. Suppose you are asked to find the number of different selections of three items chosen from three red, two white, two blue, one black and one yellow. There are three essentially different situations:

XXX   three items all the same colour. There is only one way to get this pattern, by choosing three red items.

        Total: 1 selection.

XXY   two items the same colour and one different. There are three ways of getting two the same (red, white or blue), and whichever of these is chosen there are then four possibilities for the third item.

        Total: $3 \times 4 = 12$ selections.

XYZ   all three items different. This is just a matter of choosing three of the five possible colours,

        and there are $\binom{5}{3}$ ways to do this.

        Total: 10 selections.

So altogether there are $1 + 12 + 10 = 23$ possible selections.

The same technique can be extended without too much effort to cases in which the objects chosen must be arranged as well. If you are asked for the number of possible arrangements of three coloured items chosen from those listed above, the same three situations must be considered.

XXX can be arranged in only 1 way.

1 selection $\times$ 1 arrangement   =   1

XXY can be arranged in $\dfrac{3!}{2!}$ =   3

12 selections $\times$ 3 arrangements   =   36

XYZ can be arranged in 3!   =   6

10 selections $\times$ 6 arrangements   =   60

Thus there are 97 possible arrangements of three items chosen from this collection.

The technique of considering different situations individually can be applied in most such cases, and a further example should help you consolidate your understanding.

Example

In how many ways can four letters be chosen from the word REMEMBRANCE and then arranged to make a word?

Solution

Remember that in this type of problem, the 'words' need not be real English words - any arrangement of four letters will do. There are seven different letters and four situations: no letter occurs four times, so XXXX is impossible.

E occurs three times, so XXXY offers $1 \times 6 = 6$ choices. Each of these can be arranged in $\dfrac{4!}{3!} = 4$ ways, so there are 24 possible arrangements of this form.

E, R and M each occur at least twice, so XXYY is possible in $\dbinom{3}{2} = 3$ ways. These letters can be arranged in $\dfrac{4!}{(2! \; 2!)} = 6$ ways, giving 18 arrangements altogether.

The pattern XXYZ can be achieved in $\dbinom{3}{1} \times \dbinom{6}{2} = 45$ ways, each of which can be arranged in $4! \div 2! = 12$ ways. There are thus 540 arrangements for this pattern.

Finally, XYZW, with four different letters, presents $\dbinom{7}{4} = 35$ choices, each of which has $4! \; = 24$ arrangements, giving 840 arrangements altogether.

Putting all these results together, there are 89 possible selections but 1422 possible arrangements of four letters taken from REMEMBRANCE.

Exercise 3D

1. In a bag there are two treacle toffees, two plain toffees, two nut toffees, two liquorice toffees, and two mint toffees. If I am invited to take three toffees, how many different choices are possible?

2. How many four-letter words are there in PHOTOGRAPHY?

3. Helen is playing a game involving coloured counters. In her hand she has five red counters, three green, two blue, two yellow and one black. How many different sets of four counters can Helen make up?

4. How many different patterns can Helen make from four of these counters arranged in a square?

5. Elmer is planning his homework schedule. He has five lots of homework to do, each lasting an hour, and four three-hour evenings in which to do them. In how many different ways could he allocate subjects to evenings, if the order within any given evening is not important?

3.5   Simple probability

The methods of the previous sections can be applied to simple (and not-so-simple) problems in probability. Where a situation has many possible outcomes, you may recall, all equally likely, the probability of 'success' (however that is defined) is given by

$$\text{Probability} = \frac{\text{number of successful outcomes}}{\text{total number of outcomes}}$$

If you want to find the probability of getting a prime number with one throw of a dodecahedral (12-faced) die, for example, you note that there are five prime-number outcomes and 12 possible outcomes altogether, so that the probability is 5/12. The extension of this idea to enumeration is not difficult, as the following examples show.

Example

From a class of 21 students, four have to be selected as stewards for Open Evening. If the selection is made at random, what is the probability that the four selected include both Lee and Geoffrey?

Solution

Total number of selections $= \binom{21}{4}$.

Selections including L and G $= \binom{19}{2}$

because all that remains is to choose the other two.

Probability of including L and G

$$= \binom{19}{2} \div \binom{21}{4} = 171 \div 5985 = \frac{1}{35}.$$

Example

If ten girls are arranged at random in a line, what is the probability that Susan is directly between Chantal and Yolanda?

Solution

Total possible arrangements = 10! If CSY is treated as one block, there are 8! arrangements with these three girls together, and the block itself can be either CSY or YSC. So the probability is

$$(2 \times 8!) \div 10! = \frac{1}{45}$$

Example

In the game of bridge, a Yarborough is a hand of thirteen cards in which there is no card higher than 9. What is the probability of being dealt such a hand? (Ace counts high.)

Solution

Total number of hands $=\binom{52}{13}$, because you receive 13 of the 52 cards and the order of receiving them is not important. Number of Yarborough hands $=\binom{32}{13}$, because there are 32 cards ranked 9 and under (four each of 2, 3, 4, 5, 6, 7, 8 and 9). Probability of Yarborough $=\binom{32}{13}\div\binom{52}{13}\approx 0.000\ 547$.

Can you solve this last problem by an alternative method, not involving enumeration?

Exercise 3E

1. If four students are chosen at random from a class of ten males and eight females, what is the probability that they are all male?

2. If eleven different cars are parked in a random order in adjacent parking bays, what is the probability that the Metro is next to the Fiesta?

3. Coming home from a ten-day Scout Camp, Pete announces that it rained on only three days. On this information alone, what is the probability that the first and last days were both dry?

4. In bridge, what is the probability that your 13-card hand includes all four aces?

5. Still in bridge, what is the probability that each of the four players receives a hand consisting of cards all of one suit? (Such an occurrence is reported in the press every couple of years.)

3.6   Subsets

"Don't take all those magazines," says Mother, "but you can have some of them if you want." Assuming there are nine magazines, all different, how many different selections of "some of them" are possible?

One way of solving this problem is to work step by step through the different numbers of magazines you might take.

With one magazine, there are $\binom{9}{1}=9$ selections.

With two magazines, there are $\binom{9}{2}=36$ selections.

With three magazines, there are $\binom{9}{3} = 84$ selections.

With eight magazines, there are $\binom{9}{8} = 9$ selections.

What are the missing values?

If you add all these answers together, you will find they add up to 510, so there are 510 ways of taking some (but not all) of the magazines.

But there is another simpler way to solve the problem. For each of the nine magazines in turn, you have a simple choices: take it or leave it - 2 possibilities. And the choice for each magazine is independent, so there are $2 \times 2 \times 2 \times \ldots \times 2 = 2^9$ possibilities altogether. But two of these apparent possibilities are not actually possible at all: you cannot take all nine magazines (because Mother told you not to), nor can you leave all nine (because then you would not even have 'some'). So in the end there are $2^9 - 2 = 510$ different selections of some but not all of nine magazines.

Another problem using a similar technique is to decide how many different amounts of money you could make from (say) five £1 coins, four 20p coins, three 5p coins and four 2p coins. Notice that the numbers and values of the coins have been carefully chosen so that the same total cannot be made in more than one way.

You can take 0, 1, 2, 3, 4 or 5 £1 coins $\Rightarrow$ 6 choices.

You can take 0, 1, 2, 3 or 4   20p coins $\Rightarrow$ 5 choices.

You can take 0, 1, 2 or 3   5p coins $\Rightarrow$ 4 choices.

You can take 0, 1, 2, 3 or 4   2p coins $\Rightarrow$ 5 choices.

So there are $6 \times 5 \times 4 \times 5 = 600$ different choices, giving 599 different amounts of money if £0.00 is excluded.

A third problem involving a twist on this method is to find the number of ways in which eight children can be split into two teams (not necessarily equal) for a game. As in Section 3.3, it is enough to choose one of the two teams, and there are $2^8 - 2$ ways to do this because the first team must include some but not all of the children. But this figure counts each split twice: once when team ABCDE is chosen, for example, and again when the choice is FGH. So the true number of allocations is $\left(2^8 - 2\right) \div 2 = 127$.

Exercise 3F

1. How many non-empty subsets has the set {red, orange, yellow, green, blue, violet}?

2. How many different committees could be formed from a club with 25 members if no member may sit on more than one committee?

3. In how many ways can nine books be split into two piles?

4. How many factors has 720? (Hint: Write 720 as a product of prime factors, and use the method of the second example.)

5. How many words (of any length) can you make from the letters of CHEMIST?

3.7   The pigeonhole principle

The pigeonhole principle (also known as Dirichlet's box principle) was mentioned in Chapter 1, and can now be considered in slightly greater depth. It states, you will recall, that if n objects have to be placed in m pigeonholes, where $n > m$, there must be at least one pigeonhole with more than one object in it. An extended version similarly asserts that if $n > km$, there must be at least one pigeonhole with more than k objects in it.

When the principle is applied to the solution of problems, the 'pigeonholes' are often mathematical ideas rather than real objects. The principle is most often used to prove assertions, as in the examples following.

Example

Prove that if five points are marked within an equilateral triangle of side 2 cm, two of them are within 1 cm of one another.

Solution

Imagine the triangle divided into four equal smaller triangles, as shown in the diagram. There are then five objects (points) to be allocated to four pigeonholes (small triangles), and by the pigeonhole principle there must be one triangle with more than one point in it. But the small triangles have side 1 cm, so any two points within such a triangle are within 1 cm of one another.

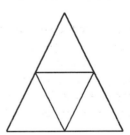

Example

Prove that among any group of six people, there are either three who all know one another or three who are all strangers to one another.

Solution

Suppose the people are A, B, C, D, E and F. Now A either knows or does not know each of the other five; if those remaining have to be put into two pigeonholes ('A knows' and 'A doesn't know') then one or other pigeonhole must contain at least three of B, C, D, E and F.

Suppose it is the 'A knows' pigeonhole, and that A knows B, C and D. Now either two of these know one another, in which case they make with A a set of three who all know one another (✓ assertion proved) or none of them know one another, in which case B, C and D are a set of three mutual strangers (✓).

Alternatively, if the 'A doesn't know' pigeonhole is the one with at least three members, suppose A doesn't know D, E or F. Then either all three of these know one another, in which case they are the three required (✓) or at least two of them are strangers, in which case they with A are a set of three mutual strangers (✓).

Either way, the assertion is proved.

* Exercise 3G

The questions in this exercise are more demanding than those normally set in A Level examinations, but provide an excellent training in mathematical thinking.

1. Show that if 51 points are marked within a square of side 7 cm, it is possible to draw a circle with radius 1 cm that contains at least three of the points.

2. Prove that in any set of ten whole numbers between 1 and 20 (inclusive), there are two whose highest common factor is greater than 1.

3. A point (x, y) in Cartesian geometry is called a lattice point if x and y are whole numbers. Prove that if five lattice points are chosen, they include two whose midpoint is also a lattice point.

4. Show that any set of n whole numbers includes a non-empty subset (which may be the whole set) whose sum is divisible by n.

5. If 17 points are joined to one another by straight lines (to form the graph K_{17}), and every line is either red, white or blue, prove that there is at least one triangle which has three sides all of the same colour.

3.8　Inclusion and exclusion

You have already seen a number of enumeration problems solved by a 'back-door' approach, in which the solution involves calculating an answer more than the real answer, and then taking some away. The tennis example in Section 3.1 is such a problem, as is Question 6 in Exercise 3C. In this section, the 'Inclusion-exclusion principle' is developed a little further.

The principle itself can be expressed in various ways. Its clearest expression is probably in set notation, where $n(A)$ represents the number of elements in the set A and Σ represents a summation.

The principle states that

$$n(A_1 \cup A_2 \cup ... \cup A_r) = \sum n(A_i) - \sum n(A_i \cap A_j) + \sum n(A_i \cap A_j \cap A_k)$$
$$- ... \pm n(A_1 \cap A_2 \cap ... \cap A_r)$$

where the signs continue alternately + and − to the end.

Example

In a large group of children, there are 23 who hate cabbage and 14 who hate semolina. If 6 children hate both cabbage and semolina, how many hate at least one of them?

Solution

Using the obvious notation,

$$n(C \cup S) = [n(C) + n(S)] - n(C \cap S)$$

$$= [23 + 14] - 6$$

$$= 31$$

This agrees with the common-sense answer, because the six children who hate both have been included in the 23 and again in the 14, and so are counted twice unless 6 is subtracted from the total.

Example

A batch of cars went for their M.O.T. test. 17 cars had faulty lights, 21 had faulty tyres, and 16 had faulty steering. 9 cars failed on lights and tyres, 12 failed on tyres and steering, and 8 failed on steering and lights. 4 cars failed on all three points.

How many cars failed the test altogether?

Solution

$$n(L \cup T \cup S) = [n(L) + n(T) + n(S)] -$$
$$[n(L \cap T) + n(T \cap S) + n(S \cap L)] + n(L \cap T \cap S)$$
$$= (17 + 21 + 16) - (9 + 12 + 8) + 4$$
$$= 29.$$

The principle can be applied to other problems too, where the set connection is less obvious.

Example

In a class there are 23 children, including four pairs of twins. In how many ways can five children be chosen if they must not include both of any pair?

Solution

There are $\binom{23}{5} = 33\ 649$ choices altogether.

Among these, there are $\begin{pmatrix} 21 \\ 3 \end{pmatrix} = 1330$ choices which include the A

twins and three other children; similarly, there are 1330 choices including the B twins, 1330 with the C twins, and 1330 with the D twins.

There are $\begin{pmatrix} 19 \\ 1 \end{pmatrix} = 19$ choices including both the A twins and the B

twins with one other child, and similarly for each of the six ways of including two pairs of twins.

So using the inclusion-exclusion principle, the number of permissible choices is

$$33\ 649 - 4 \times 1330 + 6 \times 19 = 28\ 443.$$

Exercise 3H

1. How many hands of seven cards which contain neither an ace nor a spade can be dealt from an ordinary pack?

2. In a Sixth Form tutor group of 18 students, 6 are taking French, 5 are taking German, and 4 are taking Spanish. 3 students are taking at least two of these languages, and 2 students are taking all three. How many are taking no languages at all?

3. On an exam paper there are five questions in Section A and five in Section B. Candidates must answer four questions, including at least one question from each section. In how many ways can such a choice be made?

4. In a political debating society there are 12 Conservatives, 14 Labour supporters, and 9 Liberal Democrats. In how many ways can a committee of 5 members be chosen if it must include at least one supporter of each party?

5. A tennis club has 16 male and 21 female members, including six married couples. In how many ways can the club select a team of three men and three women, not including both halves of any married couple?

3.9   Unequal division

If Elsie, Lacie and Tillie have nine treacle toffees to share between them, in how many ways can they do it? If the shares have to be equal, there is obviously just one way - each girl must have three toffees, and since the toffees are all the same it doesn't matter which three.

But what if the shares need not all be the same size?

How many divisions are possible then?

The answer depends on whether or not you insist that every girl should have at least one toffee. If you do, then imagine the toffees set out in a line with two pencils somewhere in the line to separate the toffees into three groups.

The first group - the one on the left - will be Elsie's share, the middle group will be Lacie's, and the last group will be for Tillie. Clearly, each placing of the pencils will give a different division of the toffees.

So how many placings of the two pencils are there? The nine toffees create eight gaps between them, and two of these gaps must be chosen for the pencils. This can be done in $\binom{8}{2} = 28$ ways, so there are 28 ways to share out the toffees with each girl getting at least one.

Activity 7

Check this result by listing all the possible divisions.

If there is no restriction on the sharing - if zero shares are allowed - there are clearly more than 28 possibilities. The same model can be used, however, and the problem now is simply to arrange nine toffees and two pencils in a line. The two pencils might be next to each other (in which case Lacie gets no toffees), or at one end (in which case Elsie or Tillie goes without). The number of ways of arranging 11 objects where 9 are the same (toffees) and 2 are the same (pencils) is just $11! \div (9! \times 2!) = 55$, so this is the number of possible divisions in this case.

These principles can be applied to other cases of unequal division.

If n identical objects have to be shared (not necessarily equally) between k different groups, the number of possible divisions is

$$\binom{n-1}{k-1}$$ if every group must have at least one object, or

$$\binom{n+k-1}{k-1}$$ if there is no such restriction.

Example

The baker sells white rolls, brown rolls, sesame rolls and fruity rolls. If I go to buy a dozen rolls, how many different selections could I bring home?

Solution

A 'baker's dozen' is traditionally 13, so the problem is to allocate 13 rolls to 4 groups, each group representing one kind of roll. There is no minimum number for any group - I don't have to take any fruity rolls at all if I don't want them - so the second formula is the one that applies. There are $\binom{16}{3} = 560$ possibilities.

Example

How many solutions in positive integers has the equation

$x + y + z = 10$? (An equation whose solutions must be positive integers is sometimes called a Diophantine equation after the Greek mathematician *Diophantus*, who studied such equations in Egypt about 1750 years ago.)

Solution

This problem needs ten 'units' to be split into three groups, where each group must contain at least one (to give a positive integer).

Using the first formula, there are $\binom{9}{2} = 36$ ways to do this, so the

equation has 36 different solutions.

Activity 8

Consider how to extend the methods above to answer the following.

(a) In how many ways can three children share twelve chocolate eclairs if each child must get at least two?

(b) An examination paper contains four questions, each worth a maximum of 10 marks. How many ways are there of scoring a total of 20 marks for the paper?

Exercise 3I

1. A test paper contains four questions, each worth a maximum of 10 marks. How many ways are there of scoring a total of 10 marks for the whole paper?

2. A forester goes out to buy ten saplings for a copse. If she wants a mixture of ash, birch and holly saplings, how many different combinations might she consider?

3. A sweet manufacturer sells 'lucky bags' each containing a mixture of chews, liquorice whirls, and fruit drops. If there are twenty sweets altogether, and at least one of each kind, how many different bags can be produced?

4. Rovers scored eight goals in their first home game last season. In how many ways could these goals have been shared among the eleven members of the team?

5. After a party, twelve students travel back to their college in three cars. If the identities of the individual students are ignored, in how many different ways could they be distributed between the cars?

* 3.10   Partitions

How many addition sums are there, using only positive integers, whose answer is 10?

This is not quite the same as the previous examples, because the problem does not specify the number of terms. What is more, it is not clear from the wording of the problem whether $6+4$ is to be considered the same as, or different from, $4+6$.

Suppose first of all that the order of terms is considered important, so that $6+4$ is different from $4+6$. Then

- with two terms, there are 9 possible sums from $1+9$ to $9+1$,
- with three terms, there are 36 sums, as in the third example above,
- with four terms, there are $\binom{9}{3} = 84$ sums,

and so on.

If all these results are added together, the total number of sums turns out to be 511.

Check this result by working out the other values for yourself.

There is an easier way, though. Any sum totalling ten can be rewritten in crude 'Roman numerals' with ten Is and one or more + signs. For example, $5+1+4$ could be written

$$I\ I\ I\ I\ I+I+I\ I\ I\ I$$

If the Is are written first they leave nine gaps in which a + might be put, and at each gap there are two choices (+ or no +). So there are $2\times2\times\ ...\ \times2 = 2^9$ choices for the row as a whole, giving 2^9-1 possible sums since the choice of no + signs at all is not allowed. The answer is thus 511 by this method too.

You might think that disregarding the order of terms, treating $6+4$ and $4+6$ as the same, would make the problem easier. In fact the opposite is the case: although the results obtained are smaller numbers, it is much more difficult to calculate them. The young Indian mathematician *Srinivasa Ramanujan* made considerable progress towards a solution of this problem in the early twentieth century, but there is still no general formula for unordered partitions comprehensible to most people.

Activity 9

There are six unordered partitions of 5, not including 5 itself:

$4+1 \quad 3+2 \quad 3+1+1 \quad 2+2+1 \quad 2+1+1+1 \quad 1+1+1+1+1$

By careful listing, find the number of unordered partitions of each of the whole numbers from 2 to 10.

If you can find a general formula linking the results you get, and equally valid for numbers beyond 10, you will almost certainly get your name in the mathematical history books!

* 3.11   Derangements

Four men go to a pub, and each of them hangs his coat on a peg near the door. After half a dozen drinks or so, the men get their coats ready to leave, but each of them takes another's coat. In how many ways is this possible?

This is a typical 'derangement' problem, involving a rearrangement in which none of the objects keeps its original place. For small numbers, derangement problems can be solved in various ways.

Method 1

List all the possible derangements. There can't possibly be more than 24 of them, because that is the total number of rearrangements of four coats, including some in which at least one man has the right coat.

In this case, if the original order is ABCD, the possible derangements are BADC, BCDA, BDAC, CADB, CDAB, CDBA, DABC, DCAB and DCBA - nine of them in all.

Method 2

Use the inclusion-exclusion principle.

Derangements $\quad = 24$ (total rearrangements)

$\qquad -4 \times 6$ (one coat correct)

$\qquad +6 \times 2$ (two coats correct)

$\qquad -4 \times 1$ (three coats correct)

$\qquad +1 \times 1$ (four coats correct)

$\quad = 9$ as before.

Activity 10

Use either of these two methods to complete the following table.

Number of objects	1	2	3	4	5
Derangements	0	1		9	

Your answer for five objects probably suggests that the numbers involved are getting quite large for these fairly crude methods, and that something more sophisticated is needed. The best way to continue the table, in fact, turns out to be a recurrence relation.

Let d_n represent the number of derangements of n objects.

Consider one of these objects in particular - object A, say. In any derangement, A must take the place previously occupied by another object - call it B - and the object B can be chosen in $(n-1)$ ways.

Now there are two possibilities:

(i)   B takes the place previously occupied by A, and the other $(n-2)$ objects are deranged among themselves: this can happen in d_{n-2} ways; or

(ii) a different object - object C, say - takes the place previously occupied by A. Every such derangement corresponds exactly to a derangement of the $(n-1)$ objects without A, where C takes the place previously occupied by B, and there are d_{n-1} derangements of this kind.

There are therefore $(n-1)$ ways of choosing B, and for any given B there are $(d_{n-1}+d_{n-2})$ possible derangements, so by multiplication we have the recurrence relation

$$d_n = (n-1)(d_{n-1}+d_{n-2})$$

(Recurrence relations, also called difference equations, are explained more fully in Chapters 14 and 15.)

Now you know already that $d_1 = 0$, $d_2 = 1$, $d_3 = 2$, $d_4 = 9$ and $d_5 = 44$, and you can check by substitution that these values do indeed satisfy the recurrence relation. For example, the relation gives $d_5 = 4 \times (d_4 + d_3) = 4 \times (9+2) = 44$ as expected.

The recurrence relation can be used to calculate d_n for any given value of n, though when n is large it is a nuisance having to proceed step by step rather than directly to a result. Unfortunately, the recurrence relation cannot be solved by the usual methods explained in Chapter 15, so that no explicit formula can be given here.

Activity 11

Use the recurrence relation to find the value of d_n for values of n up to 10.

One extension, however, may be of interest. The problem as commonly posed concerns an incompetent secretary who types letters to n different people, each with its own envelope, but then puts the letters into the envelopes entirely at random. What is the probability that no letter is in its correct envelope?

As n increases, will the probability tend to 1, or to 0, or to some other value? Discuss this with other students before reading on.

Since there are d_n possible derangements of n letters, and $n!$ possible arrangements altogether, the probability of a derangement is $d_n \div n!$.

Activity 12

Use your previous answers to help you complete the following table.

n	1	2	3	4	5	6	7	8	9	10
d_n										
$n!$										
Probability										

What do you notice? Can you predict (at least approximately) the value of d_{100}?

3.12 Miscellaneous Exercises

1. There are eight runners in the 100 metres final. Assuming there are no dead heats, in how many different orders might they finish?

2. From my collection of twelve teddy bears, I must choose just three to take on holiday with me. How many different choices might I make?

3. How many solutions in positive integers has the equation $a+b+c+d+e=10$?

4. In how many ways can the letters of the word MATHEMATICS be arranged?

5. A tennis club has 12 male and 15 female members. In how many ways can two 'mixed doubles' pairs be selected to represent the club in a tournament?

6. In how many ways can Anne, Bert, Cora, Dawn, Ella and Fred sit on a row of chairs if Bert and Fred must be kept separate?

7. A club with 24 members must elect a chairman, a secretary and a treasurer. In how many ways can it do this?

8. In how many ways can eight members of a chess club be paired off to play one against another?

9. How many factors has the number 7200?

10. What is the probability that a five-card poker hand, dealt from a single pack, contains at least one card of each suit?

11. In how many ways can eight rooks be placed on a chessboard so that no two of them attack one another? (Two rooks attack one another if they are on the same row or the same column.)

12. How many four-letter words can be made using letters from MISSISSIPPI?

13. How many numbers between 1 and 2000 inclusive will not divide by any prime number less than 10?

14. How many squares (of all sizes) are there on an ordinary chessboard? And how many rectangles?

15. Prove that in any set of eleven whole numbers there are two whose difference divides exactly by 10.

16. If four married couples sit down at a round table, men and women alternately, in how many different ways can they sit if no husband is next to his own wife?

17. If five identical cubical dice are thrown at the same time, how many different results are possible? (A 'result' for this purposes is a set of scores such as 2-2-3-5-6; the order of the dice does not matter.)

*18. Using only 'silver' coins (i.e. 50p, 20p, 10p and 5p), in how many ways can you give change for £1?

*19. On an examination paper, Question 1 carries 20 marks and Questions 2, 3 and 4 carry 10 marks each. If all four questions are to be answered, show that there are 1111 ways in which a candidate can score exactly 30 marks.

*20. If $S(n, k)$ (called a Stirling number of the second kind) denotes the number of ways in which a set with n elements can be partitioned into k non-empty subsets, show that
$S(n,k)=k\,S(n-1,k)+S(n-1,k-1)$. Hence or otherwise, find $S(6, 3)$.

21. There are $n\,(\geq 4)$ identical items in a row and they are split into four groups each consisting of one or more item. The first few in the row will form the first group, the next few will from the second group and so on. For example 0 0 I 0 0 0 I 0 I 0 0 0 0.

By considering where the breaks between groups must be drawn, explain why the row can be split into four such groups in $\binom{n-1}{3}$ ways.

Use this result to answer the following questions.

(a) In how many different ways can 30 identical sweets be shared out amongst four children so that each child gets at least one sweet?

(b) By first giving each of the children 5 sweets, or otherwise, calculate the number of ways in which 50 identical sweets can be shared out amongst four children so that each child gets at least 6 sweets.

(c) By first taking a sweet from each child, or otherwise, calculate the number of ways in which 30 identical sweets can be shared out amongst four children, where this time some children may get no sweets.

(AEB)

4 INEQUALITIES

Objectives

After studying this chapter you should

- be able to manipulate simple inequalities;
- be able to identify regions defined by inequality constraints;
- be able to use arithmetic and geometric means;
- be able to use inequalities in problem solving.

4.0   Introduction

Since the origin of mankind the concept of one quantity being greater than, equal to or less than, another must have been present. Human greed and 'survival of the fittest' imply an understanding of inequality, and even as long ago as 250 BC, Archimedes was able to state the inequality

$$3\frac{10}{71} < \pi < 3\frac{10}{70}.$$

Nowadays we tend to take inequalities for granted, but the concept of **inequality** is just as fundamental as that of **equality**.

You certainly meet inequalities throughout life, though often without too much thought. For example, in the United Kingdom the temperature T°C is usually in the range

$$-15 < T < 30$$

and it would be extremely cold or hot if the temperature was outside this range. In fact, animal life can exist only in the narrow band of temperature defined by

$$-60 < T < 60.$$

It will be assumed that you are familiar with a basic understanding of the use of inequalities $<$, $\leq$, $>$ and $\geq$, and that you have already met the graphical illustration of simple inequalities. You will cover this ground again, but experience with this using Cartesian coordinates and some competence in algebraic manipulation would be very helpful.

Activity 1

Find and prove an inequality relationship for π.

4.1 Fundamentals

The concept of 'greater than' or 'less than' enables numbers to be ordered, and represented on, for example, a number line.

The time line opposite gives a time scale for some important events.

You can also use inequalities for other quantities. For example, the speed of a small car will normally lie within the limits

$$-15\text{mph} < \text{speed} < 120\text{mph}.$$

Before looking at more inequality relationships the definition must be clarified. Writing $x > y$ simply means that $x - y$ is a positive number: the other inequalities $<$, $\geq$, $\leq$ can be defined in a similar way. Using this definition, together with the fact that the sum, product and quotient of two positive numbers are all positive, you can prove various inequality relationships.

Example

Show that

(a) if $u > v$ and $x > y$ then $u + x > v + y$;

(b) if $x > y$ and k is a positive number, then $kx > ky$.

Solution

(a) If $u > v$ and $x > y$ then this simply means that $u - v$ and $x - y$ are both positive numbers: hence their sum

$$u - v \text{ and } x - y$$

is also positive. But this can be rewritten as

$$(u + x) - (v + y).$$

Since this difference is a positive number you can deduce that

$$u + x > v + y$$

as required.

(b) If $x > y$ then this simply means that $x - y$ is a positive number. Since k is also positive you can deduce that the product $k(x - y)$ is positive. Therefore

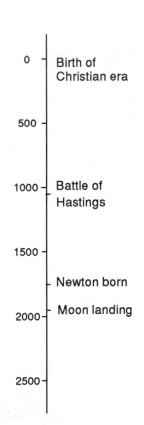

0	Birth of Christian era
500	
1000	Battle of Hastings
1500	
	Newton born
2000	Moon landing
2500	

$$kx - ky = k(x - y)$$

is a positive number, which means that $kx > ky$ as required.

Activity 2

What happens to property (b) when k is a **negative** number?

What happens to subtraction of inequalities? For example, if $u > v$ and $x > y$ then is it always true that $u - v > x - y$?

Can you take square roots through an inequality? i.e. If $a^2 > b^2$ then is it necessarily true that $a > b$?

Investigate these questions with simple illustrations.

In what follows you will need to solve and interpret inequalities. These inequalities will usually be **linear** (that means not involving powers of x, etc), but you will first see how to solve more complex inequalities. The procedure is illustrated in the following example.

Example

Find the values of x which satisfy the inequality

$$x^2 + 7 < 3x + 5.$$

Solution

You can rewrite the inequality as

$$x^2 + 7 - (3x + 5) < 0$$

$$x^2 - 3x + 2 < 0$$

$$(x - 2)(x - 1) < 0.$$

Since the complete expression is required to be negative, this means that one bracket must be positive and the other negative.

This will be the case when $1 < x < 2$.

Exercise 4A

1. Prove that if $x > y > 0$, then $\dfrac{1}{y} > \dfrac{1}{x}$.

2. Prove that if $a^2 > b^2$, where a and b are positive numbers, then $a > b$.

3. Find the values of x for which $8 - x \geq 5x - 4$.

4. Find in each case the set of real values of x for which

   (a) $3(x - 1) \geq x + 1$

   * (b) $\dfrac{3}{(x-1)} \geq \dfrac{1}{(x+1)}$

5. Find the set of values of x for which
$$x^2 - 5x + 6 \geq 2.$$

4.2   Graphs of inequalities

In the last section it was shown that inequalities can be solved algebraically; however, it is often more instructive to use a graphical approach.

Consider the previous example in which you want to find values of x which satisfy

$$x^2 + 7 < 3x + 5.$$

Another approach is to draw the graphs of

$$y_1 = x^2 + 7, \quad y_2 = 3x + 5$$

and note when $y_1 < y_2$. This is illustrated in the graph opposite.

Between the points of intersection, A and B, $y_2 > y_1$. Solving the equation $y_1 = y_2$ gives

$$x^2 + 7 = 3x + 5$$

$$\Rightarrow \quad x^2 - 3x + 2 = 0$$

$$\Rightarrow \quad (x - 2)(x - 1) = 0$$

$$\Rightarrow \quad x = 1 \text{ or } 2$$

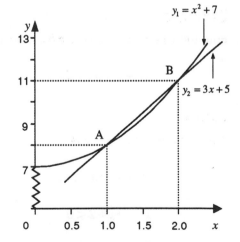

giving, as before, the solution $1 < x < 2$.

You will find a graphical approach particularly helpful when dealing with inequalities in two variables.

Example

Find the region which satisfies $2x + y > 1$.

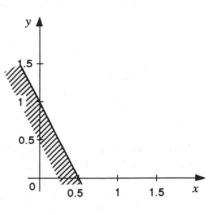

Solution

The boundary of the required region is found by solving the **equality**

$$2x + y = 1.$$

This is shown in the diagram opposite

The inequality will be satisfied by all points on one side of the line. To identify which side, you can test the point $(0, 0)$ - this does not satisfy the inequality, so the region to the right of the line is the solution. The **excluded** region is on the **shaded** side of the line.

Just as you can solve simultaneous equations, you can tackle simultaneous inequalities. For example, suppose you require values of x and y which satisfy

$$2x + y > 1$$

and $\qquad x + 2y > 1.$

You have already solved the first inequality, and if you add on the graph of the second inequality, you obtain the region as shown in this diagram.

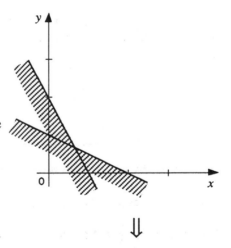

Combining the two inequalities gives the solution region as shown opposite.

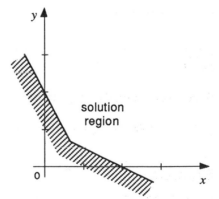

Example

Find the region which satisfies all of the following inequalities.

$$x + y > 2$$
$$3x + y > 3$$
$$x + 3y > 3$$

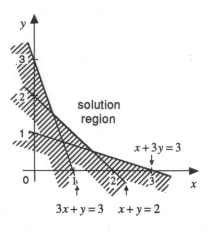

Solution

As before, the graph of the three inequalities is first drawn and the region in which **all** three are satisfied is noted.

Note that if you had wanted to solve

$$x + y < 2$$
$$3x + y > 3$$
$$x + 3y > 3$$

then the solution would have been the triangular region completely bounded by the three lines; in general the word **finite** will be used for such bounded regions.

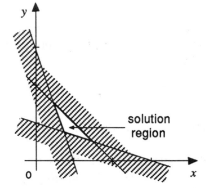

Activity 3

Write down **three** different linear equations of the form

$$ax + by = c.$$

Which three inequalities are satisfied in the finite region formed by these lines?

Suppose you now have **four** linear inequalities in x and y to be satisfied.

What regions might they define?

The following example illustrates some of the possibilities.

Example

In each case find the solution region.

(a) $x + y > 1, \ y - x < 1, \ 2y - x > 0, \ 2x + 3y < 6$

(b) $x + y > 1, \ y - x < 1, \ 2y - x < 0, \ 2x + 3y > 6$

(c) $x + y < 1, \ y - x > 1, \ 2y - x < 0, \ 2x + 3y > 6$

Solution

First graph $x+y=1$, $y-x=1$, $2y-x=0$ and $2x+3y=6$, and then in each case identify the appropriate region.

There is no region which satisfies (c).

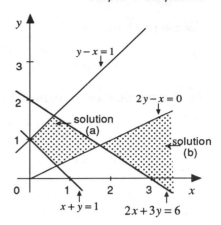

How many finite regions are formed by the intersection of four lines?

Exercise 4B

1. Solve graphically the inequalities

    (a) $(2-3x)(1+x) \le 0$

    (b) $x^2 \le 2x+8$.

2. Solve

    $y \ge 0$, $x+y \le 2$ and $y-2x < 2$.

3. Find the solution set for

    $x+y < 1$ and $3x+2y < 6$.

4. Find the region which satisfies

    $x+y \ge 2$
    $x+4y \le 4$
    $y > -1$.

5. Is the region satisfying

    $x+y > 1$, $3x+2y < 12$, $y-x < 2$, $2y-x > 1$

    finite?

    $2x+3y = 6$

4.3   Classical inequalities

You are probably familiar with the arithmetic mean (often called the average) of a set of positive numbers. The **arithmetic mean** is defined for positive numbers x_1, x_2, ..., x_n by

$$A = \frac{x_1 + x_2 + \ ... \ + x_n}{n}$$

So, for example, if $x_1 = 5$, $x_2 = 6$, $x_3 = 10$, then

$$A = \frac{(5+6+10)}{3} = 7.$$

There are many other ways of defining a mean; for example, the **geometric mean** is defined as

$$G = \left(x_1 x_2 \ ... \ x_n\right)^{1/n}$$

For the previous example,

$$G = (5 \times 6 \times 10)^{\frac{1}{3}} = (300)^{\frac{1}{3}} \approx 6.69$$

The **harmonic mean** is defined by

$$\boxed{\frac{1}{H} = \frac{1}{n}\left(\frac{1}{x_1} + \frac{1}{x_2} + \ldots + \frac{1}{x_n}\right)}$$

So, again with $x_1 = 5$, $x_2 = 6$, and $x_3 = 10$,

$$\frac{1}{H} = \frac{1}{3}\left(\frac{1}{5} + \frac{1}{6} + \frac{1}{10}\right) = \frac{1}{3} \times \frac{7}{15}$$

giving

$$H = \frac{45}{7} \approx 6.43$$

Activity 4

For varying positive numbers x_1, x_2, x_3, find the arithmetic, geometric and harmonic means. What inequality can you conjecture which relates to these three means?

If you have tried a variety of data in Activity 4, you will have realised that the geometric and harmonic means give less emphasis to more extreme numbers. For example, given the numbers 1, 5 and 9,

$$A = 5, \ G = 3.56, \ H = 2.29,$$

whereas for numbers 1, 5 and 102,

$$A = 36, \ G = 7.99, \ H = 2.48.$$

Whilst the arithmetic mean has changed from 5 to 36, the geometric mean has only doubled, and the harmonic mean has hardly changed at all!

In most calculations for mean values the **arithmetic mean** is used, but not always.

One criterion which any mean must satisfy is that, when all the numbers are equal, i.e. when $x_1 = x_2 = \ldots = x_n(= a)$ say, then the mean must equal a.

For example,

$$A = \frac{a+a+ \ldots +a}{n} = \frac{na}{n} = a$$

$$G = (a\,a\,a\,\ldots\,a)^{\frac{1}{n}} = \left(a^n\right)^{\frac{1}{n}} = a.$$

Similarly, $H = a$ when all the numbers are equal.

Activity 5

Define a new mean of n positive numbers $x_1, x_2, \ldots, x_n$ and investigate its properties.

In Activity 4 you might have realised that

$$A \geq G \geq H$$

(equality only occurring when all the numbers are equal). The first inequality will be proved for any two positive numbers, x_1 and x_2.

Given the inequality

$$(x_1 - x_2)^2 \geq 0$$

then equality can occur only when $x_1 = x_2$.

This inequality can be rewritten as

$$x_1^2 - 2x_1x_2 + x_2^2 \geq 0$$

or $\qquad x_1^2 + 2x_1x_2 + x_2^2 \geq 4x_1x_2$ (adding $4x_1x_2$ to each side)

giving $\qquad \dfrac{(x_1 + x_2)^2}{4} \geq x_1x_2.$

Taking the positive square root of both sides, which was justified in Question 2 of Exercise 4A,

$$\boxed{\frac{x_1 + x_2}{2} \geq \sqrt{x_1x_2}}$$

i.e. $\qquad A \geq G$

and equality only occurs when $x_1 = x_2$.

You will see how this result can be used in geometrical problems.

Example

Show that of all rectangles having a given perimeter, the square encloses the greatest area.

Solution

For a rectangle of sides a and b, the perimeter, L is given by

$$L = 2(a+b),$$

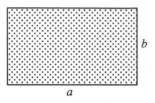

and the area, A, by

$$A = ab.$$

Using the result above, with x_1 replaced by a and x_2 replaced by b,

$$\frac{L}{4} \geq \sqrt{A}$$

or $\qquad A \leq \frac{L^2}{16}$

where equality occurs only when $a = b$. Since in this example the perimeter is fixed, the right hand side of this last inequality is constant: also equality holds if and only if $a = b$. Therefore you can deduce that the maximum value of A is $\dfrac{L^2}{16}$ and that it is only obtained for the square.

In fact, the inequalities

$$A \geq G \geq H$$

hold for any set of positive numbers, x_1, x_2, ..., x_n, but the result is not easy to prove, and requires, for example, the use of a mathematical process called induction.

The result in the example above illustrates what is called an **isoperimetric inequality**; you will see more of these in the next section.

Exercise 4C

1. Find the arithmetic, geometric and harmonic means for the following sets of numbers, and check that the inequality $A \geq G \geq H$ holds in each case.

   (a) 1, 2, 3, 4;

   (b) 0.1, 2, 3, 4.9;

   (c) 0.1, 2, 3, 100;

   (d) 0.001, 2, 3, 1000;

   (e) 0.001, 0.002, 1000, 2000.

2. Prove that $G \geq H$ for any two positive numbers x_1 and x_2.

*3. By taking functions $\frac{1}{x^2}$ and x^2 as numbers in the arithmetic/geometric mean inequality, find the least value of

$$y = \frac{1+x^4}{x^2}.$$

*4. Show that the surface area, S, of a closed cylinder of volume V can be written as

$$S = 2\pi r^2 + \frac{2V}{r}.$$

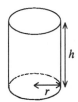

Writing

$$S = 2\pi\left(r^2 + \frac{2V}{2\pi r}\right)$$

and using the arithmetic and geometric means inequality for the three numbers

$$r^2, \frac{V}{2\pi r}, \frac{V}{2\pi r}$$

show that

$$\frac{S}{6\pi} \geq \left(\frac{V^2}{4\pi^2}\right)^{\frac{1}{3}}.$$

When does equality occur? What relationship does this give between h and r?

4.4   Isoperimetric inequalities

In the last section there was an example of an isoperimetric inequality. You will look at a more general result (first known to the Greeks in about 2000 BC) and at some further special cases.

According to legend, Princess Dido was fleeing from the tyranny of her brother and, with her followers, set sail from Greece across the sea. Having arrived at Carthage, she managed to obtain a grudging concession from the local native chief to the effect that

> 'she could have as much land as could be encompassed by an ox's skin.'

Of course, the natives expected her to kill the biggest ox she could find and use its skin to claim her land - but her followers were very astute, advising her to cut the skin to make as many thin strands as possible and to join them together to form one long length to mark the perimeter of her land. Her only problem then was in deciding what shape this perimeter should be to enclose the maximum area.

What do you think is the best shape?

In mathematical terms, the search is for the shape which maximises the area A inside a given perimeter of length L.

area A

perimeter length L

Example

For a given perimeter length, say 12 cm, find the area enclosed by

(a)   a square;

(b)   a circle;

(c)   an equilateral triangle.

Solution

(a)   For $L = 12$, each side is of length 3 cm

and $A = 3^2 = 9 \text{ cm}^2$.

(b)   For $L = 12$, assume the radius is a, giving

$$12 = 2\pi a \Rightarrow a = \frac{6}{\pi}$$

and $A = \pi a^2 = \pi \left(\frac{6}{\pi}\right)^2 = \frac{36}{\pi} \approx 11.46 \text{ cm}^2$

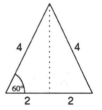

(c)   Again, for $L = 12$, each side is of length 4 cm, and

$$A = \frac{1}{2} \times 4 \times 4 \sin 60 = 4\sqrt{3} \approx 6.93 \text{ cm}^2.$$

So, for the particular problem of a perimeter length of 12 cm, of the three shapes chosen the circle gives the largest area - but can there be another shape which gives a larger one? You can make some progress by looking more carefully at the circle in the general case of perimeter L.

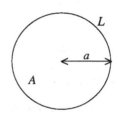

Now $L = 2\pi a$

and $A = \pi a^2 = \pi \left(\frac{L}{2\pi}\right)^2 = \frac{L^2}{4\pi}$

So, for **any** circle

$$\frac{4\pi A}{L^2} = 1.$$

The **Isoperimetric Quotient Number** (I.Q.) of any closed curve is defined as

$$\boxed{\text{I. Q.} = \frac{4\pi A}{L^2}}$$

For the circle, you see that I.Q. = 1. In the basic problem you have been trying to find the shape which gives a maximum value to A for a fixed value of L. In terms of the I.Q. number, you want to find the shape which gives the maximum value to the I.Q. number. But, for a circle, the value of the I.Q. number is 1, so if the optimum shape is a circle, then the inequality

$$\boxed{\text{I.Q.} \leq 1}$$

holds for all plane shapes, and equality occurs **only** for the circle.

Note that, since the I.Q. is the ratio of an area to the square of a length, it is non-dimensional, i.e. a number requiring no units.

Example

Find the I.Q. number for a square of side a.

Solution

$$L = 4a, \quad A = a^2,$$

and

$$\text{I.Q.} = 4\pi \times \frac{a^2}{(4a)^2} = \frac{\pi}{4} \approx 0.785.$$

Activity 6

Find I.Q. numbers of various shapes and check that, in each case, the inequality I.Q. ≤ 1 holds.

A complete proof is beyond the scope of this present work (and, in fact, involves high level mathematics). It is surprising that such a simple result, known to the Greeks, could not be proved until the late 19th Century, and even then required sophisticated mathematics. You can, though, verify the result for all regular polygons as will be shown.

Consider a regular polygon of n sides. The angle subtended by each side at the centre is

$$\frac{360}{n} \text{ degrees} \quad \text{or} \quad \frac{2\pi}{n} \text{ radians.}$$

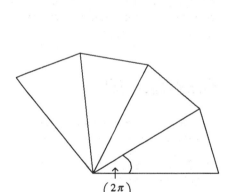

You will work in **radians** in what follows. If each side is of length a, the area of each triangle is given by

$$\frac{1}{2} \times a \times \frac{a}{2} \times \frac{1}{\tan\left(\dfrac{\pi}{n}\right)} = \frac{a^2}{4\tan\left(\dfrac{\pi}{n}\right)}$$

The total area, $\quad A = \dfrac{na^2}{4\tan\left(\dfrac{\pi}{n}\right)}, \quad$ and $\quad L = na,$

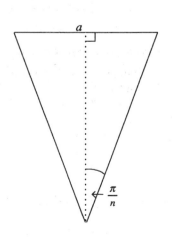

so

$$\text{I.Q.} = 4\pi \times \left(\frac{na^2}{4\tan\left(\dfrac{\pi}{n}\right)} \right) \times \frac{1}{(na)^2}$$

i.e.

$$\text{I.Q.} = \frac{\left(\dfrac{\pi}{n}\right)}{\tan\left(\dfrac{\pi}{n}\right)}$$

Activity 7

Use your calculator to find the limit of $\left(\dfrac{x}{\tan x}\right)$ as $x \to 0$.

Now you can write $I.Q. = \dfrac{x}{\tan x}$ where $x = \dfrac{\pi}{n}$.

As $n \to \infty$, the polygon becomes, in the limit, a circle, and you have seen that $I.Q. = 1$, as expected. Note that, for all values of the positive integer n except $n = 1$, $\tan\left(\dfrac{\pi}{n}\right) > \left(\dfrac{\pi}{n}\right)$ - use your calculator to check some of the values. Hence, for any regular ploygon

$$I.Q. \leq 1,$$

and you can see that the larger n becomes, the closer the I.Q. comes to 1, fitting in with the fact that the I.Q. for a circle is 1.

Finally, it should be noted that Princess Dido did not live happily ever after. Having been outwitted by her, the native leader promptly fell in love with her. As she did not reciprocate his feelings, she burnt herself on a funeral pyre in order to escape a fate worse than death!

Exercise 4D

1. For a given perimeter length of 12 cm, find the total area enclosed by the rectangle with sides
   (a) 3 cm and 3 cm
   (b) 2 cm and 4 cm
   (c) 1 cm and 5 cm

2. Find the I.Q. numbers for the following shapes:
   (a) equilateral triangle;
   (b) regular hexagon;
   (c) rectangle with sides in the ratio 1 : 2.

3. For a rectangle with sides in the ratio
   $$1 : k \ \ (k \geq 1),$$
   find an expression for the I.Q. number. What value of k gives:
   (a) maximum value
   (b) minimum value
   for the I.Q. number?

4. What is the volume, V, of the sphere which is enclosed by a surface area of 12 cm^2?

5. What is the volume, V, of the cube which is enclosed by a surface area of 12 cm^2?

6. For a given surface area, S, what closed three-dimensional shape do you think gives a maximum volume?

4.5   Miscellaneous Exercises

1. Obtain the sets of values of x for which

    (a) $2x > \dfrac{1}{x}$

    (b) $\dfrac{1}{x+1} > \dfrac{x}{3+x}$.

2. Find the range of values of x for which

    $4x^2 - 12x + 5 < 0$.

3. Find the ranges of values of x such that

    $x > \dfrac{2}{x-1}$.

4. Find the set of values of x for which

    $\dfrac{x(x+2)}{x-3} < x+1$.

5. Find the solution set of the pair of inequalities

    $x + y < 1$

    $2x + 5y < 10$.

6. Is the region defined by

    $2x - 3y \le 6$

    $x + y \le 4$

    finite?

7. Find the region satisfied by

    $x + y \le 4$

    $2x - 3y \le 6$

    $3x - y \ge -3$

    $x \le 2$.

8. Prove that $A \ge H$ for any two positive numbers. When does equality occur?

9. Find the I.Q. number for the shape illustrated below, where k is a positive constant.

    What value of k gives a maximum I.Q. value?

10. Find the I.Q. number for a variety of triangles, including an equilateral triangle. What do you deduce about the I.Q. numbers for triangles?

*11. For three-dimensional closed shapes, the isosurface area quotient number is defined as

    $$\text{I.Q.} = \frac{6\sqrt{\pi}\,V}{S^{\frac{3}{2}}}$$

    where V is the volume enclosed by a total surface area S. Find the I.Q. for a variety of three-dimensional shapes. Can you find an inequality satisfied by all closed shapes in three dimensions?

5 LINEAR PROGRAMMING

Objectives

After studying this chapter you should

*   be able to formulate linear programming problems from contextual problems;

*   be able to identify feasible regions for linear programming problems;

*   be able to find solutions to linear programming problems using graphical means;

*   be able to apply the simplex method using slack variables;

*   understand the simplex tableau procedure.

5.0   Introduction

The methods of linear programming were originally developed between 1945 and 1955 by American mathematicians to solve problems arising in industry and economic planning. Many such problems involve constraints on the size of the workforce, the quantities of raw materials available, the number of machines available and so on. The problems that will be solved usually have two variables in them and can be solved graphically, but problems occurring in industry have many more variables and have to be solved by computer. For example, in oil refineries, problems arise with hundreds of variables and tens of thousands of constraints.

Another application is in determining the best diet for farm animals such as pigs. In order to maximise the profit a pig farmer needs to ensure that the pigs are fed appropriate food and sufficient quantities of it to produce lean meat. The pigs require a daily allocation of carbohydrate, protein, amino acids, minerals and vitamins. Each involves various components. For example, the mineral content includes calcium, phosphorus, salt, potassium, iron, magnesium, zinc, copper, manganese, iodine, and selenium. All these dietary constituents should be present, in correct amounts.

A statistician at Exeter University has devised a computer program for use by farmers and companies producing animal feeds which enables them to provide the right diet for pigs at various stages of development, such as the weaning, growing and finishing stages. The program involves 20 variables and 10 equations!

Undoubtably linear programming is one of the most widespread methods used to solve management and economic problems, and has been applied in a wide variety of situations and contexts.

5.1 Formation of linear programming problems

You are now in a position to use your knowledge of inequalities from the previous chapter to illustrate **linear programming** with the following case study.

Suppose a manufacturer of printed circuits has a stock of

200 resistors, 120 transistors and 150 capacitors

and is required to produce two types of circuits.

Type A requires 20 resistors, 10 transistors and 10 capacitors.

Type B requires 10 resistors, 20 transistors and 30 capacitors.

If the profit on type A circuits is £5 and that on type B circuits is £12, how many of each circuit should be produced in order to maximise the profit?

You will not actually solve this problem yet, but show how it can be formulated as a linear programming problem. There are three vital stages in the formulation, namely

(a) What are the unknowns?

(b) What are the constraints?

(c) What is the profit/cost to be maximised/minimised?

For this problem,

(a) **What are the unknowns?**

Clearly the number of type A and type B circuits produced; so we define

x = number of type A circuits produced

y = number of type B circuits produced

(b) **What are the constraints?**

There are constraints associated with the total number of resistors, transistors and capacitors available.

Resistors Since each type A requires 20 resistors, and each type B requires 10 resistors, then

$$20x + 10y \le 200,$$

as there is a total of 200 resistors available.

Transistors   Similarly

$$10x + 20y \leq 120$$

Capacitors   Similarly

$$10x + 30y \leq 150$$

Finally you must state the obvious (but nevertheless important) inequalities

$$x \geq 0, \; y \geq 0.$$

(c) **What is the profit?**

Since each type A gives £5 profit and each type B gives £12 profit, the total profit is £P, where

$$P = 5x + 12y.$$

You can now summarise the problem as:

maximise $P = 5x + 12y$

subject to $20x + 10y \leq 200$

$\qquad\qquad 10x + 20y \leq 120$

$\qquad\qquad 10x + 30y \leq 150$

$\qquad\qquad x \geq 0$

$\qquad\qquad y \geq 0.$

This is called a **linear** programming problem since both the objective function P and the constraints are all linear in x and y.

In this particular example you should be aware that x and y can only be integers since it is not sensible to consider fractions of a printed circuit. In all linear programming problems you need to consider if the variables are integers.

Activity 1   Feasible solutions

Show that $x = 5, y = 3$ satisfies all the constraints.

Find the associated profit for this solution, and compare this profit with other possible solutions.

At this stage, you will not continue with finding the actual solutions but will concentrate on further practice in formulating problems of this type.

The key stage is the first one, namely that of identifying the unknowns; so you must carefully read the problem through in order to identify the basic unknowns. Once you have done this successfully, it should be straight forward to express both the constraints and the profit function in terms of the unknowns.

A further condition to note in many of the problems is that the unknowns must be positive **integers**.

Example

A small firm builds two types of garden shed.

Type A requires 2 hours of machine time and 5 hours of craftsman time.

Type B requires 3 hours of machine time and 5 hours of craftsman time.

Each day there are 30 hours of machine time available and 60 hours of craftsman time. The profit on each type A shed is £60 and on each type B shed is £84.

Formulate the appropriate linear programming problem.

Solution

(a) **Unknowns**

Define

x = number of Type A sheds produced each day,

y = number of Type B sheds produced each day.

(b) **Constraints**

Machine time: $2x + 3y \leq 30$

Craftsman time: $5x + 5y \leq 60$

and $x \geq 0, y \geq 0$

(c) **Profit**

$P = 60x + 84y$

So, in summary, the linear programming problem is

maximise $P = 60x + 84y$

subject to $2x + 3y \leq 30$

$x + y \leq 12$

$x \geq 0$

$y \geq 0$

Exercise 5A

1. Ann and Margaret run a small business in which they work together making blouses and skirts.

   Each blouse takes 1 hour of Ann's time together with 1 hour of Margaret's time. Each skirt involves Ann for 1 hour and Margaret for half an hour. Ann has 7 hours available each day and Margaret has 5 hours each day.

   They could just make blouses or they could just make skirts or they could make some of each.

   Their first thought was to make the same number of each. But they get £8 profit on a blouse and only £6 on a skirt.

   (a) Formulate the problem as a linear programming problem.

   (b) Find three solutions which satisfy the constraints.

2. A distribution firm has to transport 1200 packages using large vans which can take 200 packages each and small vans which can take 80 packages each. The cost of running each large van is £40 and of each small van is £20. Not more than £300 is to be spent on the job. The number of large vans must not exceed the number of small vans.

   Formulate this problem as a linear programming problem given that the objective is to **minimise** costs.

3. A firm manufactures wood screws and metal screws. All the screws have to pass through a threading machine and a slotting machine. A box of wood screws requires 3 minutes on the slotting machine and 2 minutes on the threading machine. A box of metal screws requires 2 minutes on the slotting machine and 8 minutes on the threading machine. In a week, each machine is available for 60 hours.

   There is a profit of £10 per box on wood screws and £17 per box on metal screws.

   Formulate this problem as a linear programming problem given that the objective is to **maximise** profit.

4. A factory employs unskilled workers earning £135 per week and skilled workers earning £270 per week. It is required to keep the weekly wage bill below £24 300.

   The machines require a minimum of 110 operators, of whom at least 40 must be skilled. Union regulations require that the number of skilled workers should be at least half the number of unskilled workers.

   If x is the number of unskilled workers and y the number of skilled workers, write down all the constraints to be satisfied by x and y.

5.2 Graphical solution

In the previous section you worked through problems that led to a linear programming problem in which a **linear** function of x and y is to be maximised (or minimised) subject to a number of **linear** inequalities to be satisfied.

Fortunately problems of this type with just two variables can easily be solved using a graphical method. The method will first be illustrated using the example from the text in Section 5.1. This resulted in the linear programming problem

maximise $P = 5x + 12y$

subject to $20x + 10y \le 200$

$$10x + 20y \le 120$$
$$10x + 30y \le 150$$
$$x \ge 0$$
$$y \ge 0$$

You can illustrate the **feasible** (i.e. allowable) region by graphing all the inequalities and shading out the regions not allowed . This is illustrated in the figure below.

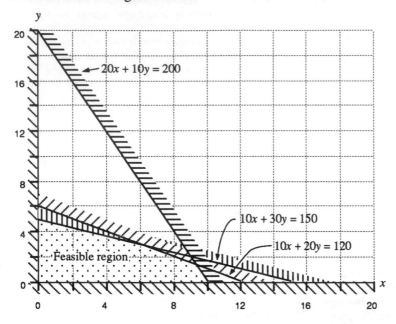

Magnifying the feasible region, you can look at the family of straight lines defined by

$$C = 5x + 12y$$

where C takes various values.

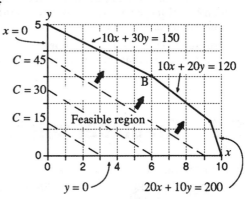

The figure shows, for example, the lines defined by

$$C = 15, \ C = 30, \text{ and } C = 45.$$

On each of these lines any point gives the same profit.

Activity 2

Check that the points

$$x = 1, \ y = \frac{25}{12}$$
$$x = 2, \ y = \frac{5}{3}$$
$$x = 4, \ y = \frac{5}{6}$$

each lie on the line defined by $C = 30$. What profit does each of these points give?

Where is the point representing maximum profit ?

As the profit line moves to the right, the profit increases and so the maximum profit corresponds to the last point touched as the profit line moves out of the feasible region. This is the point B, the intersection of

$$10x + 30y = 150 \text{ and } 10x + 20y = 120$$

Solving these equations gives $10y = 30$, i.e. $y = 3$ and $x = 6$. So maximum profit occurs at the point (6, 3) and the profit is given by

$$P = 5 \times 6 + 12 \times 3 = 66.$$

Example

A farmer has 20 hectares for growing barley and swedes. The farmer has to decide how much of each to grow. The cost per hectare for barley is £30 and for swedes is £20. The farmer has budgeted £480.

Barley requires 1 man-day per hectare and swedes require 2 man-days per hectare. There are 36 man-days available.

The profit on barley is £100 per hectare and on swedes is £120 per hectare.

Find the number of hectares of each crop the farmer should sow to maximise profits.

Solution

The problem is formulated as a linear programming problem:

(a) **Unknowns**

$x =$ number of hectares of barley

$y =$ number of hectares of swedes

(b) **Constraints**

Land $x + y \leq 20$

Cost $30x + 20y \leq 480$

Manpower $x + 2y \leq 36$

(c) **Profit**

$$P = 100x + 120y$$

To summarise, maximise $P = 100x + 120y$

$$\text{subject to } x + y \leq 20$$
$$30x + 20y \leq 480$$
$$x + 2y \leq 36$$
$$x \geq 0$$
$$y \geq 0$$

The feasible region is identified by the region enclosed by the five inequalities, as shown below. The profit lines are given by

$$C = 100x + 120y$$

and again you can see that C increases as the line (shown dotted) moves to the right. Continuing in this way, the maximum profit will occur at the intersection of

$$x + 2y = 36 \text{ and } x + y = 20$$

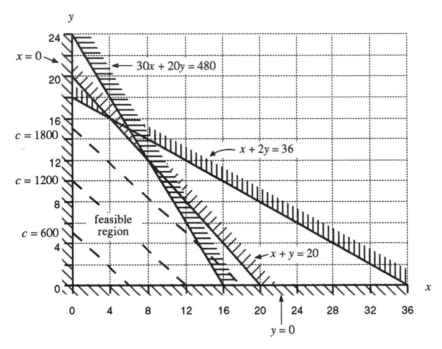

At this point $x = 4$ and $y = 16$, and the corresponding maximum profit is given by

$$P = 100 \times 4 + 120 \times 16 = 2320$$

The farmer should sow 4 hectares with barley and 16 with swedes.

Exercise 5B

1. Solve the linear programming problem defined in Question 1 of Exercise 5A.

2. Solve the linear programming problem defined in Question 2 of Exercise 5A.

3. A camp site for caravans and tents has an area of 1800m² and is subject to the following regulations:

    The number of caravans must not exceed 6.

    Reckoning on 4 persons per caravan and 3 per tent, the total number of persons must not exceed 48.

At least 200 m² must be available for each caravan and 90 m² for each tent.

The nightly charges are £2 for a caravan and £1 for a tent.

Find the greatest possible nightly takings.

How many caravans and tents should be admitted if the site owner wants to make the maximum profit and have

(a) as many caravans as possible,

(b) as many tents as possible?

4. The annual subscription for a tennis club is £20 for adults and £8 for juniors. The club needs to raise at least £800 in subscriptions to cover its expenses.

The total number of members is restricted to 50. The number of junior members is to be between one quarter and one third of the number of adult members.

Represent the information graphically and find the numbers of adult and junior members which will bring in the largest amount of money in subscriptions.

Find also the least total membership which will satisfy the conditions.

 5. The numbers of units of vitamins A, B and C in a kilogram of foods X and Y are as follows:

Food	Vitamin A	Vitamin B	Vitamin C
X	5	2	6
Y	4	6	2

A mixture of the two foods is made which has to contain at least 20 units of vitamin A, at least 24 units of vitamin B and at least 12 units of vitamin C.

Find the smallest total amount of X and Y to satisfy these constraints.

Food Y is three times as expensive as Food X. Find the amounts of each to minimise the cost and satisfy the constraints.

5.3 Simplex method

Where will a linear programming solution always occur?

Looking back at the second example in Section 5.2, the slope of the profit line was $\frac{5}{6}$. (The slope is actually negative but it is sufficient to just consider the magnitude of the slopes of the lines.) This is more than the slope of the line $x + 2y = 36$ (namely $\frac{1}{2}$), but less than the slope of the other two lines, $x + y = 20$ (i.e. 1) and $30x + 20y = 480$ (i.e. $\frac{3}{2}$).

So the solution will occur at the intersection of the two lines with slopes $\frac{1}{2}$ and 1.

Activity 3

Check the slopes of the constraints and profit function in the first example in the text in Section 5.2.

Another point worth noting here is that the solution of a linear programming problem will occur, in general, at one of the **vertices** of the feasible region if either non-integer solutions are acceptable or the vertex in question happens to have integral coordinates.

So an alternative to the graphical method of solution would be to

(a) find all the vertices of the feasible region;

(b) find the value of the profit function at each of these vertices;

(c) choose the one which gives maximum value to the profit function.

(We have not considered the possibility of non-integer vertices, which may not make sense in terms of the original problem.)

You can see how this method works with the second example in Section 5.2. The vertices are given by

(a) 0 (0, 0)

(b) A (0, 18)

(c) B (4, 16)

(d) C (8, 12)

(e) D (16, 0)

and the corresponding profits in £ are

Point	Profit
0	0
A	2160
B	2320
C	2240
D	1600

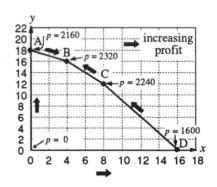

As you can see, as you move round the feasible region, the profit increases from 0 to A to B, but then decreases to C to D and back to 0.

In more complicated problems, it is helpful to introduce the idea of **slack variables**. For the problem above, with the three inequalities

$$x + y \leq 20$$
$$30x + 20y \leq 480$$
$$x + 2y \leq 36$$

three new variables are defined by

$$r = 20 - x - y$$
$$s = 480 - 30x - 20y$$
$$t = 36 - x - 2y$$

The three inequalities can now be written as

$$r \geq 0, \ s \geq 0, \ t \geq 0$$

as well as $x \geq 0$, $y \geq 0$. The variables r, s and t are called the slack variables as they represent the amount of slack between the total quantity available and how much is being used.

The importance of the slack variables is that you can now define each vertex in terms of two of the variables, $x, y, r, s,$ or t, being zero; for example, at A,

$$x = t = 0.$$

Activity 4

Complete the table below, defining each vertex

$$0 \qquad x = y = 0$$

$$A \qquad t = x = 0$$

$$B \qquad \dots\dots\dots$$

$$C \qquad \dots\dots\dots$$

$$D \qquad \dots\dots\dots$$

The procedure of increasing the profit from one vertex to the next will be again followed. Starting at the origin

$$P = 100x + 120y = 0 \text{ at } x = y = 0$$

Clearly P will increase in either direction from the origin - moving up the y axis means that x is held at zero whilst y increases. Although you know from the diagram that the next vertex reached will be A, how could you work that out without a picture? As x is being kept at zero and y is increasing the next vertex met will either be where $x = r = 0$ or where $x = s = 0$ or where $x = t = 0$. But note that

$$x = r = 0 \Rightarrow y = 20$$

$$x = s = 0 \Rightarrow y = 24$$

$$x = t = 0 \Rightarrow y = 18 \leftarrow \text{ smallest}$$

and so as y increases the first of these three points it reaches is the one where y is the smallest, namely $y = 18$ and the vertex is where $x = t = 0$ (which is the point A, thus confirming without a picture what has already been seen). You can now express the profit P in terms of x and t:

$$P = 100x + 120y$$
$$= 100x + 120\frac{(36 - x - t)}{2}$$
$$= 40x - 60t + 2160$$

How can you tell if P will increase as x increases?

So you now increase x, keeping t at zero. You will next meet a vertex where either $t = s = 0$, $t = r = 0$ or $t = y = 0$. As x is increasing the vertex will be the one of those three points where x is the smallest:

$$t = s = 0 \Rightarrow x = 6$$
$$t = r = 0 \Rightarrow x = 4 \leftarrow \text{smallest}$$
$$t = y = 0 \Rightarrow x = 36$$

So the next vertex reached is where $t = r = 0$, (which is B) and P is now expressed in terms of t and r. To do this eliminate y from the first and third of the original equations by noting that

$$2r - t = 2(20 - x - y) - (36 - x - 2y) = 4 - x$$

Hence

$$P = 40x - 60t + 2160$$
$$= 40(4 - 2r + t) - 60t + 2160$$
$$= -80r - 20t + 2320$$

How can you tell that P has reached its maximum?

Throughout the feasible region you know that $r \geq 0$ and $t \geq 0$ and so it is clear from the negative coefficients in the above expression for P that P reaches its maximum value of 2320 when r and t are both zero. This happens when $x = 4$ and $y = 16$.

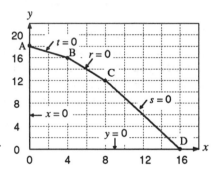

Activity 5

Now travel round the feasible region from 0 to D to C to B. At each vertex, express P in terms of the defining variables, and check that P will continue to increase until B is reached.

Although this probably looks a much more complicated way of solving linear programming problems, its real application is to problems of more than 2 variables. These cannot be solved graphically, but can be solved using a procedure using slack variables called the **simplex** method.

Exercise 5C

1. (a) Solve the linear programming problem

    maximise $P = 2x + 4y$

    subject to $x + 5y \leq 10$
    $$4x + y \leq 8$$
    $$x \geq 0$$
    $$y \geq 0$$

    by a graphical method.

    (b) Introduce slack variables r and s, and solve the problem by the simplex method.

2. (a) Determine the vertices of the feasible region for the linear programming problem

    maximise $P = x + y$

    subject to $x + 4y \leq 8$
    $$2x + 3y \leq 12$$
    $$3x + y \leq 9$$
    $$x \geq 0$$
    $$y \geq 0$$

    Hence find the solution.

    (b) Verify this solution by using the simplex method.

3. Use the simplex method to solve the linear programming problem

$$\text{maximise } P = 10x + 15y$$

$$\text{subject to } \quad 4y + 10x \leq 40$$
$$10y + 3x \leq 30$$
$$5y + 4x \leq 20$$
$$x \geq 0$$
$$y \geq 0$$

*5.4 Simplex tableau

The way this method works will be illustrated with the example.

Maximise $P = x + 2y$

subject to $x + 4y \leq 20$

$$x + y \leq 8$$
$$5x + y \leq 32$$
$$x \geq 0$$
$$y \geq 0$$

As usual introduce slack variables r, s and t defined by

$$x + 4y + r = 20$$
$$x + y + s = 8$$
$$5x + y + t = 32$$

and write the equations in the matrix form

$$P - x - 2y \qquad\qquad = 0$$
$$x + 4y + r \qquad\quad = 20$$
$$x + y \quad + s \qquad = 8$$
$$5x + y \qquad\quad + t = 32$$

$$\Rightarrow \quad
\begin{bmatrix}
1 & -1 & -2 & 0 & 0 & 0 \\
0 & 1 & 4 & 1 & 0 & 0 \\
0 & 1 & 1 & 0 & 1 & 0 \\
0 & 5 & 1 & 0 & 0 & 1
\end{bmatrix}
\begin{bmatrix}
P \\ x \\ y \\ r \\ s \\ t
\end{bmatrix}
=
\begin{bmatrix}
0 \\ 20 \\ 8 \\ 32
\end{bmatrix}$$

The **augmented matrix** with the extra right hand column will be used.

P	x	y	r	s	t			Comments
1	−1	−2	0	0	0	0		Increase x first (y could have been chosen) and compare values of x where $y=r=0$, $y=s=0$, and $y=t=0$, namely $\frac{20}{1}, \frac{8}{1}, \frac{32}{5}$*. These values are easily spotted from the matrix as the R.H. figures divided by the corresponding coefficients of x. (*This term has the smallest positive value so now manipulate the matrix to express P in terms of y and t)
0	1	4	1	0	0	20		
0	1	1	0	1	0	8		
0	5	1	0	0	1	32		
1	−1	−2	0	0	0	0		
0	1	4	1	0	0	20		
0	1	1	0	1	0	8		
0	1	$\frac{1}{5}$	0	0	$\frac{1}{5}$	$\frac{32}{5}$	$\leftarrow R_4/5$	
1	0	$-\frac{9}{5}$	0	0	$\frac{1}{5}$	$\frac{32}{5}$	$\leftarrow R_1+R_4$	From the first row express P in terms of y and t, with a positive coefficient of y. Increase y and compare the values obtained from the figures in the R.H. column divided by the corresponding coefficients of y, namely $(68/5)/(19/5)$, $(8/5)/(4/5)$*, $(32/5)/(1/5)$. (*This term has the smallest positive value (where $s=t=0$) so now manipulate the matrix to express P in terms of s and t)
0	0	$\frac{19}{5}$	1	0	$-\frac{1}{5}$	$\frac{68}{5}$	$\leftarrow R_2-R_4$	
0	0	$\frac{4}{5}$	0	1	$-\frac{1}{5}$	$\frac{8}{5}$	$\leftarrow R_3-R_4$	
0	1	$\frac{1}{5}$	0	0	$\frac{1}{5}$	$\frac{32}{5}$		
1	0	$-\frac{9}{5}$	0	0	$\frac{1}{5}$	$\frac{32}{5}$		
0	0	$\frac{19}{5}$	1	0	$-\frac{1}{5}$	$\frac{68}{5}$		
0	0	1	0	$\frac{5}{4}$	$-\frac{1}{4}$	2	$\leftarrow R_3/(\frac{4}{5})$	
0	1	$\frac{1}{5}$	0	0	$\frac{1}{5}$	$\frac{32}{5}$		
1	0	0	0	$\frac{9}{4}$	$-\frac{1}{4}$	10	$\leftarrow R_1+\frac{9}{5}R_3$	From the first row P could now be expressed in terms of s and t, with a positive coefficient of t. So now increase t and compare the values $6/(3/4)$*, $2/(-1/4)$, $6/(1/4)$ (*This term has the smallest positive value so now manipulate the matrix to express P in terms of r and s)
0	0	0	1	$-\frac{19}{4}$	$\frac{3}{4}$	6	$\leftarrow R_2-\frac{19}{5}R_3$	
0	0	1	0	$\frac{5}{4}$	$-\frac{1}{4}$	2		
0	1	0	0	$-\frac{1}{4}$	$\frac{1}{4}$	6	$\leftarrow R_4-\frac{1}{5}R_3$	
1	0	0	0	$\frac{9}{4}$	$-\frac{1}{4}$	10		
0	0	0	$\frac{4}{3}$	$-\frac{19}{3}$	1	8	$\leftarrow R_2/(\frac{3}{4})$	
0	0	1	0	$\frac{5}{4}$	$-\frac{1}{4}$	2		
0	1	0	0	$-\frac{1}{4}$	$\frac{1}{4}$	6		
1	0	0	$\frac{1}{3}$	$\frac{2}{3}$	0	12	$\leftarrow R_1+\frac{1}{4}R_2$	From the first row P could now be expressed in terms of r and s, with a negative coefficient of each, so now stop; i.e. since the top row has all positive coefficients you can see that the maximum value of P is 12 and that it is reached when $r=s=0$ (which happens when $x=4$ and $y=4$).
0	0	0	$\frac{4}{3}$	$-\frac{19}{3}$	1	8		
0	0	1	$\frac{1}{3}$	$-\frac{1}{3}$	0	4	$\leftarrow R_3+\frac{1}{4}R_2$	
0	1	0	$-\frac{1}{3}$	$\frac{4}{3}$	0	4	$\leftarrow R_4-\frac{1}{4}R_2$	

The advantage of this method is that it can be readily extended to problems with more than two variables, as shown below

Example

Maximise $P = 4x + 5y + 3z$

subject to $8x + 5y + 2z \leq 3$

$$3x + 6y + 9z \leq 2$$

$$x, y, z \geq 0$$

Solution

As usual slack variables r and s are introduced;

$$8x + 5y + 2z + r = 3$$
$$3x + 6y + 9z + s = 2$$

Now x, y, z, r, $s \geq 0$ and the simplex tableau is shown below

P	x	y	z	r	s			Comments
1	-4	-5	-3	0	0	0		Increase x initially and compare $3/8*$, $2/3$.
0	8	5	2	1	0	3		This smaller value of x occurs where $y = z = r = o$ and so now manipulate the matrix
0	3	6	9	0	1	2		to express P in terms of y, z and r.
1	-4	-5	-3	0	0	0		
0	1	$\frac{5}{8}$	$\frac{2}{8}$	$\frac{1}{8}$	0	$\frac{3}{8}$	$\leftarrow R_2/8$	
0	3	6	9	0	1	2		
1	0	$-\frac{5}{2}$	-2	$\frac{1}{2}$	0	$\frac{3}{2}$	$\leftarrow R_1 + 4R_2$	Increase y and compare $(3/8)/(5/8)$,
0	1	$\frac{5}{8}$	$\frac{2}{8}$	$\frac{1}{8}$	0	$\frac{3}{8}$		$(7/8)/(33/8)*$. This smaller value occurs when $z = r = s = o$ and so express P in terms of
0	0	$\frac{33}{8}$	$\frac{33}{4}$	$-\frac{3}{8}$	1	$\frac{7}{8}$	$\leftarrow R_3 - 3R_2$	z, r and s.
1	0	$-\frac{5}{2}$	-2	$\frac{1}{2}$	0	$\frac{3}{2}$		
0	1	$\frac{5}{8}$	$\frac{2}{8}$	$\frac{1}{8}$	0	$\frac{3}{8}$		
0	0	1	2	$-\frac{1}{11}$	$\frac{8}{33}$	$\frac{7}{33}$	$\leftarrow R_3 / \left(\frac{33}{8}\right)$	
1	0	0	3	$\frac{3}{11}$	$\frac{20}{33}$	$\frac{67}{33}$	$\leftarrow R_1 + \frac{5}{2}R_3$	The first row now has positive coefficients, showing that there is a maximum of $67/33$
0	1	0	-1	$\frac{2}{11}$	$-\frac{5}{33}$	$\frac{8}{33}$	$\leftarrow R_2 - \frac{5}{8}R_3$	when $z = r = s = o$ (which happens when $x = 8/33$, $y = 7/33$ and $z = 0$).
0	0	1	2	$-\frac{1}{11}$	$\frac{8}{33}$	$\frac{7}{33}$		

Exercise 5D

Use the simplex algorithm to solve the following problems.

1. Maximise $P = 4x + 6y$

   subject to
   $$x + y \leq 8$$
   $$7x + 4y \leq 14$$
   $$x \geq 0$$
   $$y \geq 0$$

2. Maximise $P = 10x + 12y + 8z$

   subject to
   $$2x + 2y \leq 5$$
   $$5x + 3y + 4z \leq 15$$
   $$x \geq 0$$
   $$y \geq 0$$
   $$z \geq 0$$

3. Maximise $P = 3x + 8y - 5z$

   subject to
   $$2x - 3y + z \leq 3$$
   $$2x + 5y + 6z \leq 5$$
   $$x \geq 0$$
   $$y \geq 0$$
   $$z \geq 0$$

4. Maximise $3x + 6y + 2z$

   subject to
   $$3x + 4y + 2z \leq 2$$
   $$x + 3y + 2z \leq 1$$
   $$x \geq 0$$
   $$y \geq 0$$
   $$z \geq 0$$

5.5   Miscellaneous Exercises

1. Find the solution to Question 3 of Exercise 5A.

2. A firm manufactures two types of box, each requiring the same amount of material.

   They both go through a folding machine and a stapling machine.

   Type A boxes require 4 seconds on the folding machine and 3 seconds on the stapling machine.

   Type B boxes require 2 seconds on the folding machine and 7 seconds on the stapling machine.

   Each machine is available for 1 hour.

   There is a profit of 40p on Type A boxes and 30p on Type B boxes.

   How many of each type should be made to maximise the profit?

3. A small firm which produces radios employs both skilled workers and apprentices. Its workforce must not exceed 30 people and it must make at least 360 radios per week to satisfy demand. On average a skilled worker can assemble 24 radios and an apprentice 10 radios per week.

   Union regulations state that the number of apprentices must be less than the number of skilled workers but more than half of the number of skilled workers.

   What is the greatest number of skilled workers than can be employed?

   Skilled workers are paid £300 a week, and apprentices £100 a week.

   How many of each should be employed to keep the wage bill as low as possible?

4. In laying out a car park it is decided, in the hope of making the best use of the available parking space (7200 sq.ft.), to have some spaces for small cars, the rest for large cars. For each small space 90 sq.ft. is allowed, for each large space 120 sq.ft. Every car must occupy a space of the appropriate size. It is reliably estimated that, of the cars wishing to park at any given time, the ratio of small to large will be neither less that 2:3 nor greater than 2:1.

   Find the number of spaces of each type in order to maximise the number of cars that can be parked.

*5. A contractor hiring earth moving equipment has the choice of two machines.

   Type A costs £25 per day to hire, needs one man to operate it and moves 30 tonnes of earth per day.

   Type B costs £10 per day to hire, needs four men to operate it and moves 70 tonnes of earth per day.

   The contractor can spend up to £500 per day, has a labour force of 64 men available and can use a maximum of 25 machines on the site.

   Find the maximum weight of earth that the contractor can move in one day.

6. A landscape designer has £200 to spend on planting trees and shrubs to landscape an area of 1000 m². For a tree he plans to allow 25 m² and for a shrub 10 m². Planting a tree will cost £2 and a shrub £5.

   If he plants 30 shrubs what is the maximum number of trees he can plant?

   If he plants 3 shrubs for every tree, what is the maximum number of trees he can plant?

7. A small mine works two coal seams and produces three grades of coal. It costs £10 an hour to work the upper seam, obtaining in that time 1 tonne of anthracite, 5 tonnes of best quality coal and 2 tonnes of ordinary coal. The lower seam is more expensive to work, at a cost of £15 per hour, but it yields in that time 4 tonnes of anthracite, 6 tonnes of best coal and 1 tonne of ordinary coal. Faced with just one order, for 8 tonnes of anthracite, 30 tonnes of best coal and 8 tonnes of ordinary coal each day, how many hours a day should each seam be worked so as to fill this order as cheaply as possible?

8. A cycle manufacturer produces two types of mountain-bike: a basic Model X and a Super Model Y. Model X takes 6 man-hours to make per unit, while Model Y takes 10 man-hours per unit. There is a total of 450 man-hours available per week for the manufacture of the two models.

   Due to the difference in demand for the two models, handling and marketing costs are £20 per unit for Model X, but only £10 per unit for Model Y. The total funds available for these purposes are £800 per week.

   Profits per unit for Models X and Y are £20 and £30 respectively. The objective is to maximise weekly profits by optimising the numbers of each model produced.

   (a) The weekly profit is £P. The numbers of units of Model X and Model Y produced each week are x and y. Express P in terms of x and y. Also write down inequalities representing the constraints on production.

   (b) By graphical means or by the simplex method, find the maximum obtainable profit and the numbers of each model manufactured which give this profit.

   (c) If competition forces the manufacturer to give a £5 discount on the price of Model X, resulting in a £5 reduction in profit, how are weekly profits now maximised? (AEB)

9. In order to supplement his daily diet someone wishes to take some Xtravit and some Yeastalife tablets. Their contents of iron, calcium and vitamins (in milligrams per tablet) are shown in the table.

Tablet	Iron	Calcium	Vitamin
Xtravit	6	3	2
Yeastalife	2	3	4

   (a) By taking x tablets of Xtravit and y tablets of Yeastalife the person expects to receive at least 18 milligrams of iron, 21 milligrams of calcium and 16 milligrams of vitamins. Write these conditions down as three inequalities in x and y.

   (b) In a coordinate plane illustrate the region of those points (x,y) which simultaneously satisfy $x \geq 0$, $y \geq 0$, and the three inequalities in (a).

   (c) If the Xtravit tablets cost 10p each and the Yeastalife tablets cost 5p each, how many tablets of each should the person take in order to satisfy the above requirements at the minimum cost? (AEB)

10. A maker of wooden furniture can produce three different types of furniture: sideboards, tables and chairs. Two machines are used in the production - a jigsaw and a lathe.

   The manufacture of a sideboard requires 1 hour on the jigsaw and 2 hours on the lathe; a table requires 4 hours on the jigsaw and none on the lathe; a chair requires 2 hours on the jigsaw and 8 hours on the lathe.

   The jigsaw can only operate 100 hours per week and the lathe for 40 hours per week. The profit made on a sideboard is £100, £40 on a table and £10 on a chair. In order to determine how best to use the two machines so as to maximise profits, formulate the problem as a linear programming problem, and solve it using the simplex tableau.

11. A diet-conscious housewife wishes to ensure her family's daily intake of vitamins A, B and C does not fall below certain levels, say 24 units, 30 units and 18 units, respectively. For this she relies on two fresh foods which, respectively, provide 8, 5 and 2 units of vitamins per ounce of foodstuff and 3, 6 and 9 units per ounce. If the first foodstuff costs 3p per ounce and the second only 2p per ounce, use a graphical method to find how many ounces of each foodstuff should be bought by the housewife daily in order to keep her food bill as low as possible.

6 PLANAR GRAPHS

Objectives

After studying this chapter you should

*   be able to use tests to decide whether a graph is planar;
*   be able to use an algorithm to produce a plane drawing of a planar graph;
*   know whether some special graphs are planar;
*   be able to apply the above techniques and knowledge to problems in context.

6.0   Introduction

This topic is introduced through an activity.

Activity 1

A famous problem is that of connecting each of three houses, as shown opposite, to all three services (electricity, gas and water) with no pipe/cable crossing another.

Try this problem. Four of the nine lines needed have already been put into the picture.

Investigate the problem for different numbers of houses and services.

What happens if the scene is on the surface of a sphere (which in reality it is) or on a torus (a ring doughnut!) or on a Möbius strip?

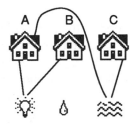

You will have found graphs which can be completed without their edges crossing and some graphs which cannot. If a graph can be drawn in the plane (on a sheet of paper) without any of its edges crossing, it is said to be **planar**.

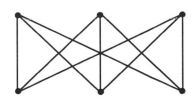

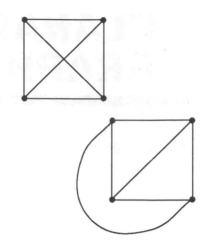

The graph shown on the right is planar, although you might not think so from the first diagram of it. The next two diagrams are of the same graph and confirm that it is planar. This diagram is called a **plane drawing** of the graph.

As you should have found in the activity above, the graph shown on the previous page, which represents the services problem, cannot in fact be drawn without crossing edges and is therefore described as **non-planar**. This has repercussions for the electronics industry, because it means that a simple circuit with six junctions and wires connecting each of three junctions to each of the other three junctions cannot be made without cross-overs. In an integrated circuit within a 'chip', this would mean two 'layers' of wires.

6.1   Plane drawings

Activity 2

Try re-drawing the two graphs shown on the right so that no edges cross.

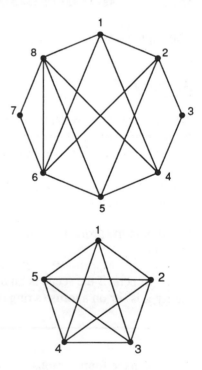

The second of the two graphs is called K_5, the complete graph with five vertices: each vertex is joined to every other one by an edge. Of course, K_6, K_7, ..., are similarly defined. Although K_5 looks simpler than the one shown above it, it is in fact non-planar, whereas the one above it is planar.

Later on it will be proved that both K_5 and another graph called $K_{3,3}$ (which is the one associated with the gas, water and electricity problem in Activity 1) are non-planar, and you will see the significance of this when you look at Kuratowski's Theorem later in this chapter.

Exercise 6A

1. Sketch the graphs of K_4 and K_6. Are they planar? For which values of n do you think K_n is planar?

2. For which values of positive integers is K_n Eulerian? For which is it semi-Eulerian?

3. There are five so-called Platonic solids with a very regular structure. Graphs based on the first three are shown below.

   Make plane drawings of each, if possible.

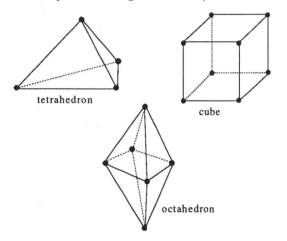

tetrahedron

cube

octahedron

4. The graph K_n can be used to represent the games played during a 'round robin' tournament in which each player plays every other player.

   How many games take place when there are

      (a)   5 players

      (b) n players?

5. Draw a graph with six vertices, labelled 1 to 6, in which two vertices are joined by an edge if, and only if, they are **co-prime** (i.e. if they have no common factor greater than 1). Is the graph planar?

6.2 Bipartite graphs

The graph associated with the activity in Section 6.1 is called a **bipartite** graph. Such graphs consist of two sets of vertices, with edges only joining vertices between sets and not within a set. The diagrams opposite are of bipartite graphs. In the second one the two sets of vertices contain three vertices and two vertices and every vertex in the first set is joined to every vertex in the second: this graph is called $K_{3,2}$ (and, of course, $K_{r,s}$ could be defined similarly for any positive integers r and s).

Exercise 6B

1. Sketch $K_{3,4}$ and $K_{4,2}$.
2. How many edges are there in general in the graph $K_{r,s}$?

3. Two opposing teams of chess players meet for some games. Show how a bipartite graph can be used to represent the games actually played. If the graph turned out to be $K_{r,s}$ what would it mean?

4. For which r and s is $K_{r,s}$ Eulerian? For which r and s is it semi-Eulerian?

6.3   A planarity algorithm

Naturally, for very complicated graphs it would be convenient to have a technique available which will tell you both whether a graph is planar and how to make a plane drawing of it.

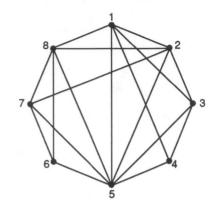

Activity 3

Using the first graph shown opposite as an example, try to develop an algorithm in order to construct a planar graph.

The algorithm described below can be applied only to graphs which have a Hamiltonian cycle; that is, where there is a cycle which includes every vertex of the graph.

The method will be illustrated by applying it to the graph shown in Activity 3, above.

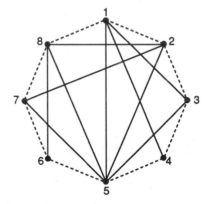

The first stage is to redraw the graph so that the Hamiltonian cycle forms a regular polygon and all edges are drawn as straight lines inside the polygon. The graph used here is already in this form, but for other graphs this stage might involve 'moving' vertices as well as edges.

The edges of the regular polygon now become part of the solution (shown dotted in the second graph).

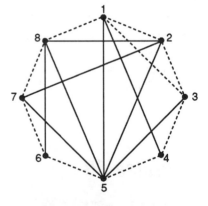

The next stage is to choose any edge, say 1 - 3, and decide whether this is to go inside or outside: let's choose inside, as illustrated in the third graph opposite.

Since 1 - 3 crosses 2 - 8, 2 - 7 and 2 - 5, all these edges must go outside as shown.

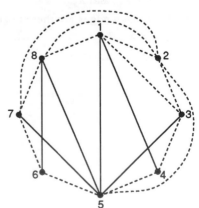

Originally edge 2 - 7 crossed 1 - 4, 1 - 5, 8 - 5 and 8 - 6 , so all these edges must now remain inside (or they would cross 2 - 7 outside).

Finally, because 1 - 4 stays inside, 3 - 5 must go outside, and since 8 - 6 stays inside, 7 - 5 must also go outside, as shown.

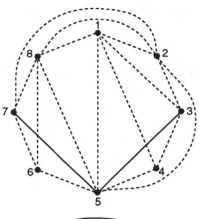

This is now a planar graph, as shown opposite, where the dotted lines have been redrawn as solid lines.

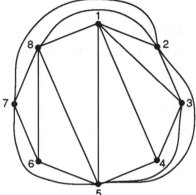

The method illustrated above can also be used to show whether or not a graph is planar. For example, consider K_5, and, as before, the regular polygon is first included as part of the solution.

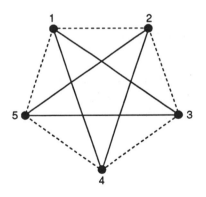

Choose an edge, say 1 - 3, which stays inside. Since this crosses 2 - 5 and 2 - 4, both of these will have to go outside as shown.

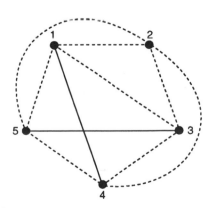

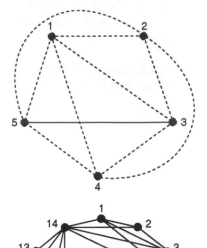

Now 2 - 5 crosses 1 - 4, so 1 - 4 must stay inside, as shown.

Finally, consider edge 3 - 5. Since it crosses 1 - 4, it must go outside; but it also crosses 2 - 4 which is already outside; so 3 - 5 must also go outside! This is a contradiction and it is concluded that the graph is non-planar.

Example

Use the planarity algorithm to find a plane drawing of the graph opposite.

Solution

The graph has a Hamiltonian cycle

$$1 - 2 - 3 - 4 - 5 - 6 - 7 - 8 - 9 - 10 - 11 - 12 - 13 - 14 - 1$$

which is part of the solution, as indicated by the dotted lines in the second graph.

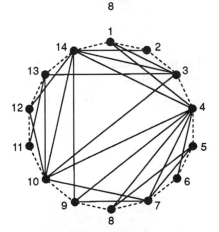

Choose any edge, say 13 - 3, and keep it in place (shown dotted). All edges that cross this line must now be put outside.

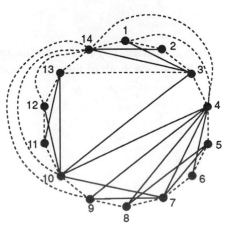

Since 1 - 4 crosses 14 - 3 and 14 - 2, they must go inside, and similarly 1 - 3 must go outside.

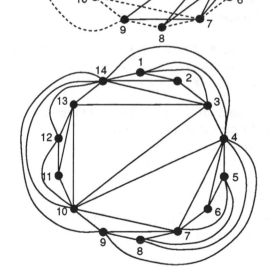

Also, 14 - 4 crosses 10 - 3 so 10 - 3 must stay inside; and since 14 - 9 is now outside, 10 - 4 and 10 - 7 must stay inside. Also 14 - 12 outside implies 13 - 10 and 13 - 11 inside, which then means that 12 - 10 must be drawn outside (as shown opposite).

Continuing in this way, 10 - 7 inside means that 9 - 4, 8 - 4, 8 - 5 all go outside; which in turn means that 9 - 7 , 7 - 4 and 6 - 4 go inside. 7 - 5 must go outside.

You now have a plane drawing of the graph as shown opposite; the lines are not dotted since this is the final solution.

Is this solution unique?

Exercise 6C

1. Make plane drawings of the following two graphs

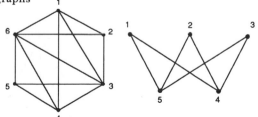

2. Show that the graphs of K_6 and $K_{3,3}$ are not planar by using the algorithm. (Note that $K_{3,3}$ must first be redrawn to form a regular polygon)

3. By first redrawing with a regular polygon, use the planarity algorithm to produce a plane drawing of the graph shown below.

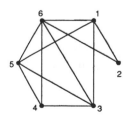

6.4 Kuratowski's Theorem

The non-planar graphs K_5 and $K_{3,3}$ seem to occur quite often.
In fact, all non-planar graphs are related to one or other of these
two graphs.

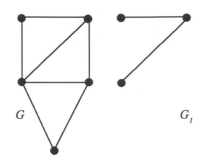

To see this you first need to recall the idea of a **subgraph**, first
introduced in Chapter 1 and define a **subdivision** of a graph.

A subgraph is simply a part of a graph, which itself is a graph.
G_1 is a subgraph of G as shown opposite.

A subdivision of a graph is the original graph with added vertices
of degree 2 along the original edges. As shown opposite, G_2 is a
subdivision of G.

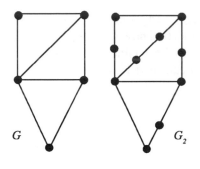

When a planar graph is subdivided it remains planar; similarly if it
is non-planar, it remains non-planar.

Kuratowski's Theorem states that a graph is planar if, and only if, it
does not contain K_5 and $K_{3,3}$, or a subdivision of K_5 or $K_{3,3}$ as a
subgraph.

This famous result was first proved by the the Polish
mathematician *Kuratowski* in 1930. The proof is beyond the scope
of this text, but it is a very important result.

The theorem will often be used to show a graph is non-planar by
finding a subgraph of it which is either K_5 or $K_{3,3}$ or a subdivision
of one of these graphs.

Example

Graph G has been redrawn, omitting edges 3 - 6 and 4 - 6. Thus G'
is a subgraph of G.

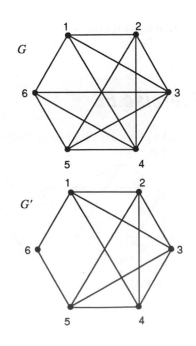

Also G' without vertex 6 is isomorphic to K_5. The addition of
vertex 6 makes G' a subdivision of K_5. So G', a subdivision of K_5,
is a subgraph of G, and therefore G is non-planar.

Exercise 6D

1. Which of these graphs are subdivisions of $K_{3,3}$ and why?

G_1 G_2 G_3

2. Use Kuratowski's Theorem to show that the following graphs are non-planar.

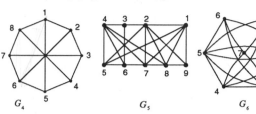 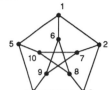

G_4 G_5 G_6

3. Show that this graph, called a Petersen graph, is non-planar.

6.5 Miscellaneous Exercises

1. Give examples of

   (a) a planar graph in which each vertex has degree 4, and

   (b) a planar graph with six vertices and a shortest cycle of length 4.

2. For which values of r, s is the complete bipartite graph $K_{r,s}$ non-planar?

3. The crossing number of a graph is the least number of points at which edges cross. What are the crossing numbers of

   (a) $K_{3,3}$

   (b) K_6

   (c) $K_{1,2}$?

4. Use Kuratowski's Theorem in order to prove that the graph G is non-planar.

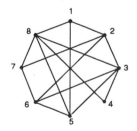

5. Show that this graph is planar.

6. The graph below represents connections in an electrical circuit. Use a planarity algorithm to decide whether or not it is possible to redraw the connections so that the graph is planar.

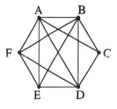

(AEB)

7 NETWORK FLOWS

Objectives

After studying this chapter you should

- be able to draw network diagrams corresponding to flow problems;

- be able to interpret networks;

- be able to find optimum flow rates in a network, subject to constraints;

- be able to use the labelling algorithm to find the maximum flow rate in a network;

- be able to interpret the analysis of a network for real life problems.

7.0   Introduction

There are many situations in life which involve flow rates; some are self-evident, such as traffic flow or the flow of oil in a pipeline; others have the same basic structure but are less obviously flow problems - e.g. movement of money between financial institutions and activity networks for building projects. In most of the problems you will meet, the objective is to maximise a flow rate, subject to certain constraints. In order to get a feel for these types of problem, try the following activity.

Activity 1

This diagram represents a road network. All vehicles enter at S and leave at T. The numbers represent the maximum flow rate in vehicles per hour in the direction from S to T. What is the maximum number of vehicles which can enter and leave the network every hour?

Which single section of road could be improved to increase the traffic flow in the network?

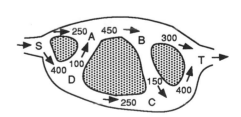

7.1 Di-graphs

The network in the previous activity can be more easily analysed when drawn as a graph, as shown opposite.

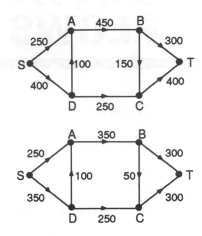

The arrows show the flow direction; consequently this is called a **directed graph** or **di-graph**. In this case the edges of the graph also have **capacities** : the maximum flow rate of vehicles per hour. The vertices S and T are called the **source** and **sink**, respectively.

You should have found that the maximum rate of flow for the network is 600. This is achieved by using each edge with flows as shown.

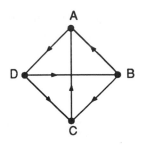

Notice that some of the edges are up to maximum capacity, namely SA, BT, DA and DC. These edges are said to be **saturated**. Also, at any vertex, other than S or T, in an obvious sense, the **inflow** equals the **outflow**.

Di-graphs for some situations show no capacities on the edges. For example, suppose you have a **tournament** in which four players each play one another. If a player A beats a player B then an arrow points from A to B. In the diagram opposite you can see that A beats D, but loses to B and C.

In what follows, the term **network** will be used to denote a directed graph with capacities.

Exercise 7A

1. The diagrams below show maximum flow capacities in network N_1, and actual intended flows in N_2.

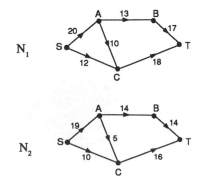

What errors have been made in constructing N_2? Draw a new network which has a maximum flow from S to T.

2. Draw a network representing the results in the tournament described by this table.

	A	B	C	D	E
A	•	X	O	X	O
B	O	•	Ⓧ	X	X
C	X	O	•	O	X
D	O	O	X	•	X
E	X	O	O	O	•

X denotes a win. O denotes a loss.
For example, the Ⓧ shows that B beats C.

7.2   Max flow - min cut

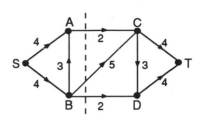

The main aim is to find the **value** of the maximum **flow** between the source and sink. You will find the concept of the **capacity** of a **cut** very useful. The network opposite illustrates a straightforward flow problem with maximum allowable flows shown on the edges.

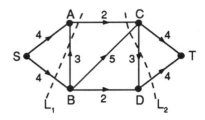

The dotted line shown in the first diagram illustrates one possible **cut**, which separates S from T. Its **capacity** is defined as the sum of the maximum allowable flows across the cut; i.e. $2+5+2=9$. There are many possible cuts across the network. Two more are shown in the second diagram. For L_1, the capacity is

$$2+0+4=6.$$

The reason for the zero is as follows: the flows in AC and SB cross the cut from left to right, whereas the flow in BA crosses from right to left. To achieve maximum flow across the cut the capacity of BA is not used.

Similarly for L_2, the capacity is given by

$$2+5+0+4=11.$$

Activity 2

For the network shown above, find all possible cuts which separate S from T, and evaluate the capacity of each cut. What is the minimum capacity of any cut?

What do you notice about the capacities?

Activity 3

Find the maximum flow for the network shown above. What do you notice about its value?

The activities above give us a clue to the max flow-min cut theorem. You should have noticed that the maximum flow found equals the cut of minimum capacity. In general,

value of any flow $\leq$ capacity of any cut

and equality occurs for maximum flow and minimum cut; this can be stated as

> maximum flow = minimum cut.

Example

For the network opposite, find the value of the maximum flow and a cut which has capacity of the same value.

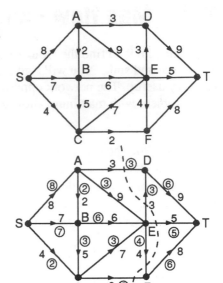

Solution

By inspection, the maximum flow has value 17; this is illustrated by the circled numbers on the network opposite. Also shown is a cut of the same capacity.

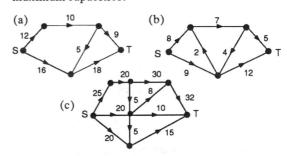

If you can find a cut and flow of the same value, can you be sure that you have found the maximum flow?

Exercise 7B

1. The network below shows maximum capacities of each edge. Draw up a table showing the values of all the cuts from A, B to C, D, E. Which is the minimum cut? Draw the network with flows which give this maximum total flow.

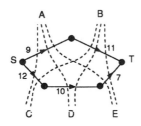

2. Find a minimum cut for each of these networks. The numbers along the edges represent maximum capacities.

    (a)

    (b)

    (c)

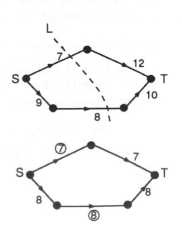

3. For each of the networks in Question 2 try to find values for the flows in each edge which give the maximum overall flow.

7.3   Finding the flow

You may have noticed that the minimum cut is coincident with edges which have a flow equal to their maximum capacity.

The diagram opposite shows a network with its allowable maximum flow along each edge. The minimum cut is marked L. It has a capacity of 15. This line cuts the edges with capacities 7 and 8. The actual maximum flow of value 15 is shown in the diagram, and it should be noted that the minimum cut only passes through edges that are saturated (or have zero flow in the opposing direction).

This information should help you to confirm maximum flows. Note that in some cases there is more than one possible pattern for the flows in the edges which give the overall maximum flow.

Activity 4

By trial and error, find the maximum possible flow for the network opposite.

Find a cut which has a capacity equal to the maximum flow (you might find it helpful to mark each edge which is satisfied by the maximum flow - the minimum cut will only cut saturated edges or edges with zero flows in the opposing direction.)

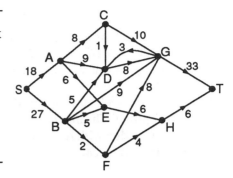

Exercise 7C

1. Find the maximum flow for each of these networks, and show the minimum cut in each case.

(a)

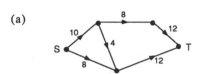

(b)

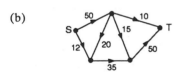

(c)

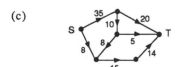

2. A network has edges with maximum capacities as shown in this table.

	S	A	B	C	D	E	T
S	•	40	40	•	•	•	
A	•	•	•	•	15	20	
B	•	•	•	45	•	•	
C	•	•	•	•	•	•	50
D	•	•	10	15	•	•	15
E	•	•	•	•	•	•	25

The letters refer to vertices of the network, where S and T are the source and sink respectively.

Draw a diagram of the network.

Find a maximum flow for the network, labelling each edge with its actual flow.

7.4 Labelling flows

So far you have no method of actually finding the maximum flow in a network, other than by intuition.

The following method describes an algorithm in which the edges are labelled with artificial flows in order to optimise the flow in each arc.

An example follows which shows the use of the **labelling algorithm**.

Example

The network opposite has a maximum flow equal to 21, shown by the cut XY. When performing the following algorithm you can stop, either when this maximum flow has been reached or when all paths from S to T become 'saturated'.

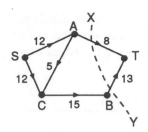

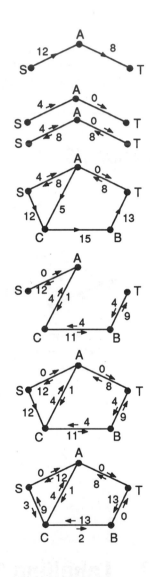

1. Note that there are three possible paths from S to T, namely SAT, SCBT, SACBT. (Note that at this stage, the directions of the flows are ignored

2. Begin with any of these, say SAT, as in the diagram opposite. The maximum flow is restricted by AT, so label each edge with its **excess capacity**, given that AT carries its marked capacity, as shown.

3. Both flows could be reduced by up to 8 (the capacity of edge AT). Show it as a potential backflow in each edge.

4. Now add this section back on to the original network as shown and choose another route, say SACBT.

   Of the possible flows, $S \rightarrow A$, $A \rightarrow C$, $C \rightarrow B$ and $B \rightarrow T$ note that the lowest is 4 and this represents the maximum flow through this path, as shown.

5. As before, each edge in the path SACBT is labelled with its excess capacity (above 4), and the reverse flows, noting that the sum of the forward and reverse flows always equals the original flow, as shown opposite. Note particularly that the excess flow in SA has now dropped to zero.

   The resulting network is shown opposite.

6. Continue by choosing a third path, say SCBT, and inserting artificial forward and backward excess flows.

   The network is shown opposite.

   There is one more route namely SCAT, but it is unnecessary to proceed with the process because the flows to T from A and B are saturated, shown by zero excess flow rates. This means that the flow can increase no further.

7. The excess flows can be subtracted from the original flows to create the actual flows or you can simply note that the back flows give the required result - but with the arrows reversed. The final result is shown opposite.

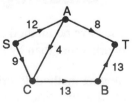

The method looks quite complicated, but after a little practice you should become quite adept at it.

The next example shows how all possible paths sometimes need to be considered.

Example

Use the labelling procedure to find the maximum flow from S to T in the network shown opposite.

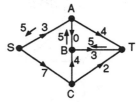

Solution

1. Possible paths SAT, SABT, SABCT, SCT, SCBT, SCBAT.

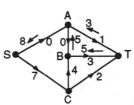

2. Start with SABT - possible to have flow of 5 units, and mark excess capacity in SA and BT and potential backflow.

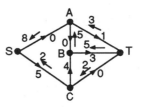

3. Now consider the path SAT - a further 3 units can flow along this path, as shown by the backflows.

4. Now consider SCT - there is a possible flow of 2 units.

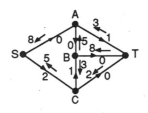

5. A further flow of 3 units is possible along the path SCBT.

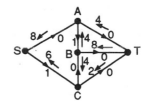

6. There is one more path to consider (since there is still excess flow along AT) namely SCBAT - this can take a further one unit (note the way the backflow and excess capacity is shown on AB).

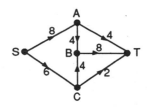

7. No more flow is possible (all areas into T have zero excess capacity), so the maximum flow, as shown opposite, has been achieved.

Exercise 7D

1. Use the labelling algorithm in order to find the maximum flow in each of these networks, given the maximum capacity of each edge.

(a)

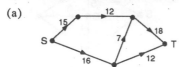

(b)

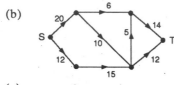

(c)

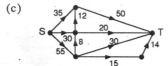

2. There are a number of road routes from town A to town B as shown in the diagram below. The numbers show the maximum flow rate of vehicles in hundreds per hour. Find the maximum flow rate of vehicles from A to B. Suggest a single road section which could be widened to improve its flow rate. How does this affect traffic flow on other sections, if the network operates to its new capacity?

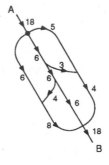

7.5   Super sources and sinks

Many networks have multiple sources and/or sinks. A road network with two sources and three sinks is shown opposite.

The problem of finding the maximum flow can be quite easily dealt with by creating a single **super source** S and a single **super sink** T.

The resulting network is as shown and the usual methods can now be applied.

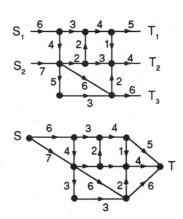

Activity 5

Add a super source and a super sink to this network, in which maximum capacities are shown, and then use the labelling algorithm to find a maximum flow through it.

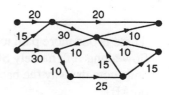

Activity 6

Investigate the traffic flow in a small section of the road network near to you, for which you could estimate maximum flows in each road.

7.6 Minimum capacities

Sometimes edges in networks also have a minimum capacity which has to be met. In the diagram opposite, for example, edge AB has a maximum capacity of 6 and a minimum of 4. The flow in this edge must be between 4 and 6 inclusive.

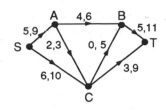

Activity 7

Find the maximum flow in the network shown above. Investigate how the max flow - min cut theorem can be adapted for this situation.

Example

Find the maximum flow for the network shown opposite.

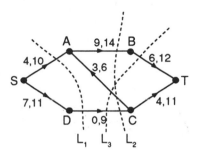

Solution

In order to find a minimum cut, the max flow - min cut theorem is adapted so that you **add** upper capacities of edges along the cut directed from S to T, but **subtract** lower capacities of edges directed from T to S.

For this network, cut L_1 has a value $10 + 9 = 19$, but cut L_2 has a value of $14 + 9 - 3 = 20$, since edge CA crosses L_2 from T to S. In fact, L_3 is the minimum cut - with a value of $12 - 3 + 9 = 18$, so you are looking for a maximum flow of 18.

If you are familiar with the labelling algorithm, here is a slightly quicker version.

1. Begin with **any** flow. The one shown opposite will do. Note that none of the upper or lower capacities of the edges has been violated. It is not the best because the flow is only 15.

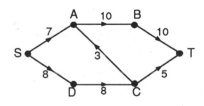

2. For each edge, insert the potential excess flow and the corresponding back flow. For example, BT carries a flow of 10 at present. It could be 2 more and it could be 4 less, since the minimum and maximum flows in the edge are 6 and 12.

   This has been done for all the edges in the network resulting in the diagram opposite.

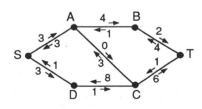

3. Now look for a path in which the flow (from S to T) can be improved. Consider SABT. The lowest excess capacity in these three edges is 2 (in BT) so the flow in each edge can be improved by this amount.

    SABT is called a **flow augmenting path** because its overall excess flow can be reduced. The reverse flow has to be increased by 2 to compensate. The path now looks like this.

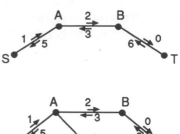

4. This can now be added to the network as shown opposite.

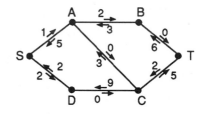

5. Now look for another path which can be augmented (improved).

    SACT cannot, since AC has a flow of zero from S to T.

    SDCABT cannot, since BT has a zero flow. The only possibility is SDCT. This can be improved by an increase of one. The result is shown here.

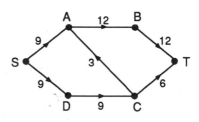

6. Now subtract the excess (S → T directed) flows from their maximum values. So, for example, AB becomes $14 - 2 = 12$. The final network - as shown opposite - has a flow of 18 as required.

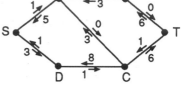

Exercise 7E

1. For the network of Activity 7 find any flow in the network and then use the labelling algorithm to find the maximum flow.

2. Which, if any, of the following networks showing upper and lower capacities of edges has a possible flow? If there is no possible flow, explain why.

    (a)

    (b)

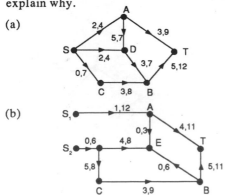

3. Find a maximum flow from S to T in this network showing upper and lower capacities.

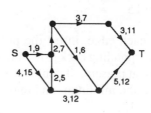

7.7 Miscellaneous Exercises

1. Find a minimum cut for each of the following networks.

   In N_1 and N_3 edges can carry the maximum capacities shown. In N_2 the minimum and maximum capacities of the edge are shown.

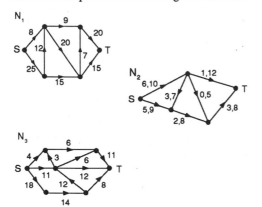

2. Use the labelling algorithm to find a maximum flow in this network, which shows maximum capacities.

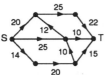

3. Which of these values of x and y gives a possible flow for the networks shown below with upper and lower capacities?

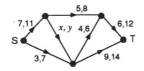

   (a) 7, 10   (b) 3, 12

   (c) 4, 5   (d) 1, 8

   For these cases find a maximum flow for the networks.

4. By creating a super source and a super sink find a maximum flow for this network which shows maximum capacities.

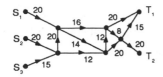

5. The table shows the daily maximum capacity of coaches between various cities (in hundreds of people).

   Draw a network to show the capacities of the routes from London through to Newcastle. A festival is taking place in Newcastle. Find the maximum number of people who can travel by coach from London for the festival. Investigate what happens when there is a strike at one of the coach stations, say Liverpool.

To From	Lon	Bir	Man	Lds	Lpl	New
London	•	40	•	20	•	•
Birmingham	•	•	10	15	12	•
Manchester	•	•	•	12	•	15
Leeds	•	•	•	•	•	30
Liverpool	•	•	7	•	•	8

6. Find a maximum traffic flow on this grid-type road system from X to Y, in which maximum flow rates are given in hundreds of vehicles per hour.

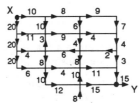

7. The following underground map shows a 'circular route' with 8 stations. Trains travel only in the direction shown. The capacities indicate the maximum number of trains per hour which can pass along each section. **At least** one train per hour must travel along each section of track. A train can carry 500 passengers. Find the maximum number of passengers which can flow from A to B.

   Note that A and B are not sources or sinks. The number of trains in the system must always remain constant.

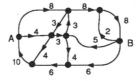

8. Consider the following network with lower and upper capacities.

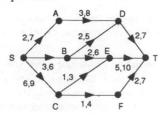

(a) Find by inspection a feasible flow from S to T whose value does not exceed 14.

(b) By looking for flow-augmenting paths (or otherwise), find a maximum flow in this network.

(c) Find a corresponding minimum cut, and check that its capacity is the same as the value of the maximum flow found in part (b).

9. Consider the following network with lower and upper capacities:

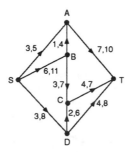

(a) Find by inspection a feasible flow from S to T whose value does not exceed 17.

(b) By looking for flow-augmenting paths (or otherwise), find a maximum flow in this network.

(c) Find a corresponding minimum cut, and check that its capacity is the same as the value of the maximum flow found in part (b).

10. Consider the following network, in which each arc is labelled with its capacity:

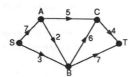

(a) Find a flow of value 9 from S to T, and draw a diagram showing the flow in each arc.

(b) Find a cut of capacity 9.

(c) What is the value of a maximum flow? (Give a brief reason for your answer.)

11. Consider the following network, where each arc is labelled with its capacity.

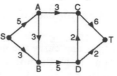

(a) Find a flow of value 7, and draw a diagram showing the flow in each arc.

(b) Find a cut of capacity 7.

(c) What is the value of a maximum flow?

(In part (c), you should give a brief reason for your answer.)

12. Verify that the max-flow min-cut theorem holds for the following network:

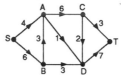

13. The network below shows the maximum and minimum flow allowed along each arc of the network.

(a) Ignoring the minimum flow constraints, find a feasible flow between S and T of value 140.

(b) Find the maximum flow, when both maximum and minimum constraints operate. Explain why your flow is a maximum flow.

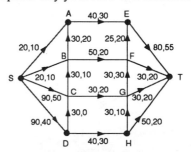

14. In the following basic network, each arc is labelled with its capacity:

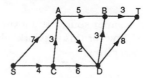

(a) Write down, or indicate on a diagram, a cut of capacity 9, separating S from T.

(b) Find a flow of value 9 from S to T, and draw a diagram showing the flow in each arc.

(c) What is the value of a maximum flow? (Give a brief reason for your answer.)

15. (a) In the following basic network, find a flow of value k from S to T, and cut with capacity k, for the same value of k.

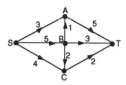

(b) Is the flow in part (a) a maximum flow? (Give a reason for your answer.)

16. The network below shows the maximum rates of flow (in vehicles per hour) between towns S, A, B, C, D and T in the direction from S to T.

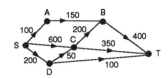

(a) By choosing a minimum cut, or otherwise, find the maximum traffic flow from S to T. Give the actual rates of flow in each of the edges BT, CT and DT when this maximum flow occurs.

(b) When a maximum flow occurs from S to T, how many of those vehicles per hour pass through C?

(c) It is decided to reduce the traffic flow through C (in the direction from S to T) to a maximum of 480 vehicles per hour. In order to maintain the same maximum flow from S to T the capacity of a single edge is to be increased. Which edge should be chosen, and by how much must its capacity be increased?

(AEB)

8 CODES IN EVERYDAY USE

After studying this chapter you should

* appreciate the role of codes in a highly technological society;

* understand how and why check digits are used;

* understand why particular designs are used for particular codes.

8.0   Introduction

Although you might not have appreciated it, many aspects of life today depend on the effective use of codes. Examples include

> **Satellite transmission**
>
> **Bar codes**
>
> **Postcodes**
>
> **Catalogue codes**
>
> **Bank codes**
>
> **Computer codes**

- the list could go on and on. Whilst the average member of the public does **not** need to know how these codes are designed or how they work, it has become a very important subject for mathematicians to study. In this chapter you will look at a number of codes used in practical situations.

8.1   Historical perspective

Although codes have now become indispensible to modern life, they are not a new invention, and our study will start with two codes which have been around for some time.

Braille

Braille is a method of writing that can be used by blind people. It was invented in 1829 by the Frenchman, *Louis Braille* (1809-52). When he was three years old he lost the sight of one eye while playing with one of his father's knives (his father was a harness maker), and soon lost his sight completely.

An earlier system for soldiers passing messages in the dark had been developed by another Frenchman, *Charles Barbier*, and this used up to **twelve** embossed dots, 6 vertical in 2 rows, as shown opposite. Each letter is made up of a pattern of raised dots which the reader can feel with his fingers. Of course, it is just as important to be able to tell when a dot is missing.

Braille revised the pattern by using a base of **six** positions, 3 vertical in 2 rows, as shown opposite.

How many different patterns exist using this system?

Activity 1

Investigate how many different patterns exist using just

(a) 1 dot   (b) 2 dots   (c) 3 dots

(d) 4 dots   (e) 5 dots   (f) 6 dots.

Check the final answer with your answer to the earlier discussion point.

The chart in Appendix 1 gives the list for the alphabet, number and punctuation. Study the chart carefully and then proceed to the next activity.

Activity 2

(a) What patterns have not been used in the given Braille chart?

(b) Can you suggest what these other patterns can be used for?

(c) Consider systems that use four or five dots as the basis rather than six. What are the advantages or disadvantages of such systems?

Morse code

This was designed in America by *Samuel Morse*, 1791 -1872, and was first used in 1844 for the telegraph line between Baltimore and Washington. Although modern technology has largely superseded the need for morse code as a form of communication, the 'SOS' code is still universally used for shipping in distress.

Activity 3

Find out the actual codes used in Morse Code. Analyse why it takes its particular form, and suggest improvements.

There are many other important historical codes, including secret codes used in the World Wars. It is the view of some that the eventual cracking of the ENIGMA code, used by the Germans in the Second World War, by the team at Bletchley, was one of the most significant factors in helping the allies to defeat Germany. This chapter, however, will deal with codes in everyday use .

8.2 Check digits

Many codes have been designed for use with new technology. These include bar codes, ISBN numbers, ASCII codes, post codes, bank sort codes; many of these modern codes employ a checking device, often referred to as a **check digit**. An example of this is that of ISBN numbers, now used universally on all new books. Each ISBN has **ten** digits made up from components, as illustrated opposite.

The check digit is designed so that any **one** error in the previous nine digits is spotted. It is calculated in the following way.

Group identifier
(up to 4 digits long) which gives information about contry of publication (UK uses '0' or '1')
↓

Check digit
↓

0   8 5 0 2 0   0 1 4   8
↑
Publisher's prefix
(one to seven digits long) uniquely identifies the publisher

↑
Title number
identifies book number in publisher's list

ISBN number

> Multiply the first nine numbers by $10, 9, 8, ..., 2$ respectively and find the sum of the resulting numbers. The check number is the smallest number that needs to be added to this total so that it is exactly divisible by 11.

For the example above, we have

$$0 \times 10 + (8 \times 9 + 5 \times 8 + 0 \times 7 + 2 \times 6 + 0 \times 5) + (0 \times 4 + 1 \times 3 + 4 \times 2)$$

$$= 135$$

so the check digit must be 8, since 143 is divisible by 11. Note that if the number 10 is needed for the check digit, the symbol X is used.

Example

Determine the check digit, a, for the following ISBN numbers:

(a)   1 869931 00 a (b)   1 7135 2272 a

Solution

(a) The number

$$1 \times 10 + (8 \times 9 + 6 \times 8 + 9 \times 7 + 9 \times 6 + 3 \times 5 + 1 \times 4) + (0 \times 3 + 0 \times 2 + 1 \times a)$$

must be divisible by 11; i.e. $266 + a$ must be divisible by 11 and $0 \le a \le 10$. Hence $a = 9$.

(b) Again, the number

$$1 \times 10 + (7 \times 9 + 1 \times 8 + 3 \times 7 + 5 \times 6) + (2 \times 5 + 2 \times 4 + 7 \times 3 + 2 \times 2 + 1 \times a)$$

must be divisible by 11; i.e. 11 divides $(175 + a)$, giving $a = 1$.

Why do ISBN numbers use a check digit of this particular form?

Activity 4   Error detection and correction

There is one error in each of these ISBN numbers. Can you correct them?

(a) 1 869932 23 8   (b) 0 7458 1078 5

Activity 5

A publisher is given a set of ISBN numbers of the form

$$1 \quad 834721 \quad m \, n \quad x$$

for $0 \le m \le 9, 0 \le n \le 9$, and x is the check digit. Design a ready reckoner or algorithm to determine the check digit for all appropriate values of m and n.

8.3   Bar code design

Bar codes are nearly universal today, being used in just about every industry. They were first suggested for automation in grocery stores in 1932 in the thesis of a Harvard Business School student, but it was not until the 1950s that the idea of a scanner installed at check-outs was conceived. It took another two decades for a combination of technology advancement and economic pressure to bring about the commercial use of bar codes and optical readers in retail trading. In 1973 the **UPC** (Universal Product Code) was adopted as a standard. In 1976 a variation known as the **EAN** (European Article Numbers) was also standardised.

Other types of bar codes, for example, **Code 3 of 9** and **Interleaved Two of Five** (ITF), have also been developed.

In the UK, the **Article Number Association** was formed to administer and promote the use of article numbering, and the association provides information packs and educational material.

8-Digit EAN

Three examples of 8-digit EAN symbols are shown opposite. These are used by large stores for their own brands. Each bar code consists of

> **left hand guard**
>
> **left hand four numbers**
>
> **centre guard**
>
> **right hand four numbers**
>
> **right hand guard.**

Looking at the code for each number, you will notice that the representation of a number is dependent on whether it is on the left or right hand side. In fact, each representation is designed using a **seven** module system. For example, a left hand side 5 is shown magnified opposite (the dashes are shown here to emphasise the seven module design - they are not actually shown on the code).

Each number has two white and two black strips of varying thickness but following the rules that

(a) the first module must be white;

(b) the last module must be black;

(c) there are in total either 3 or 5 black modules.

A convenient way of representing each number is given by using 0 (white) 1 (black) giving 0 1 1 0 0 0 1 for 5, as shown.

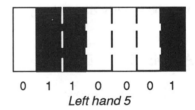

0 1 1 0 0 0 1
Left hand 5

Activity 6   Left hand codes

With the rules listed above, write down all the possible codes for left hand numbers.

Appendix 2 gives the complete set of codes for left hand numbers - called **Number Set A**. The codes for the right hand side are determined by interchanging 0 s and 1 s (i.e. white and black interchanged) - called **Number Set C**.

Why is a different code needed for right hand numbers?

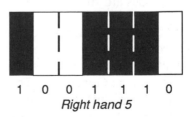

1 0 0 1 1 1 0
Right hand 5

As with ISBN numbers, these bar codes incorporate a check digit, again the last one. It is chosen so that,

$$3 \times (\text{1st} + \text{3rd} + \text{5th} + \text{7th number}) + (\text{2nd} + \text{4th} + \text{6th} + \text{8th number})$$

is exactly divisible by 10. For example, for

0033 7793

it means that

$$3 \times (0 + 3 + 7 + 9) + (0 + 3 + 7 + 3) = 3 \times 19 + 13 = 70$$

is exactly divisible by 10.

Example

Find the check digit, a, for the 8-digit EAN code

5021 421 a

Solution

The number

$$3 \times (5 + 2 + 4 + 1) + (0 + 1 + 2 + a) = 36 + 3 + a = 39 + a$$

must be exactly divisible by 10, so $a = 1$.

Activity 7   Errors

The 8-digit EAN code

5026 8020

has one error. Can you identify it?

What are the advantages of this method of determining the check digit?

13-Digit EAN

Examples of this code are found on many grocery products. Three such codes are shown opposite.

The first digit, which as you can see is not represented directly in the code, together with the second digit, indicates the country in which the article number was allocated; e.g. 50 represents the UK, 31 represents France, etc. The next five digits are issued to a particular manufacturer, and the next five identify the product. The final number is again the check digit.

All six right hand numbers are coded using **Number Set C** but the six left hand numbers are coded using a combination of **Number Sets A** and **B** (see Appendix 3) according to the first digit. For example, if the first digit is 5, then the next six digits are coded according to the Number Sets **A B B A A B**..

Using the tables in Appendix 3, can you see how Number Set B is obtained?

Activity 8

Using three As and three Bs, how many different possible combinations exist for the coding of the six left hand numbers in the code?

In fact, the first digit 0 uses the code **A A A A A A**, whereas all other first digits are coded using 3 As and 3 Bs as indicated in Appendix 4.

13-digit EAN codes use the same method as 8-digit EAN codes for determing the check digit, except that all 13 numbers are included, so that the number

$$3 \times (2\text{nd} + 4\text{th} + \ldots + 12\text{th number}) + (1\text{st} + 3\text{rd} + \ldots + 13\text{th number})$$

must be divisible by 10.

Activity 9

Check that the three 13-digit EAN codes shown earlier have correct check digits.

There are many other types of bar codes in use, some having a completely different design (e.g. library cards).

Exercise 8A

1. Design a method of coding for alphanumeric (i.e. number and letter) characters used for display on calculators.

2. Find out the code used for semaphore. Is it an efficient method of coding?

3. Marks and Spencer, who only stock their own label brands, use a special 7-digit bar code. Find out what method is used for the check digit.

8.4 Postcodes

Much of the mail in the UK is now sorted automatically. This has been made possible by the introduction of POSTCODES, which were started in 1966 and are now used throughout the UK.

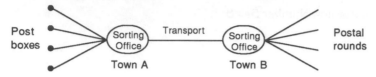

After collection, letters are sorted at the local Sorting Office into **areas** and **districts**. They are then forwarded to the appropriate Sorting Office where they are sorted again into **sectors** and **units**.

The postcode shown opposite illustrates these aspects.

Why is a mixture of numbers and letters used?

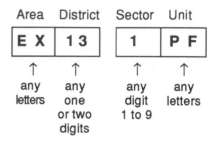

Activity 10

Keeping in mind the restrictions indicated in the postcode diagram, estimate the maximum number of units which can be defined.

In fact, there are

12 areas, 2900 districts, 9000 sectors and 2 000 000 units.

Since there are about 24 million household and business addresses in the UK, the average number of addresses per unit is given by

$$\frac{24 \times 10^6}{2 \times 10^6} = 12.$$

Why do you think the Post Office does not identify each address with a unique postcode?

Finally, it should also be noted that each postcode has to be coded (with a series of small blue dots) on the envelope to enable the automatic sorting to take place. So yet another code is used in order to make use of the first code!

Activity 11

Design a coding system, which can be put on envelopes to represent postcodes using a series of dots, to facilitate automatic sorting.

8.5   Telephone numbers

Most UK telephone numbers take the form of 10 digits as shown opposite.

The first digit is always 0, and the first digit of both the **area code** and the **local number** cannot use 0 or 1.

0	392	217113
↑	↑	↑
fixed	area code	local number

Telephone number

Activity 12

With the restrictions given above, how many unique telephone numbers exist?

There are about 25 million numbers in use in the UK but, despite your answer to Activity 3, British Telecom is in fact running out of usable numbers.

Can you suggest why?

To solve the problem of lack of codes, BT is adopting a new system of area codes introduced on 'Phoneday'

Sunday April 16th 1995.

Most local numbers will **not** change, but all area codes will have a '1' inserted after the initial '0'. For example:

0392	will become	01392
0742	will become	01742
071	will become	0171
081	will become	0181
etc.		

What advantage will this new system have?

Activity 13

List the possible disadvantages of the new system to be implemented. Consider other solutions to the problem, giving the advantages and disadvantages.

There are numerous other codes used extensively; for example

Vehicle registration numbers

Argos Catalogue numbers

ASCII codes in computing

Mariner 9 code

Cyphers

all of which have been designed to solve particular problems.

8.6 Computing codes

There are many codes used in computing, but the most commonly used code is **ASCII** (American Standard Code for Information Interchange). The code is summarised opposite. It is in ascending binary order in each section.

Character	Code
Space	010 0000
0	011 0000
1	011 0001
2	011 0010
3	011 0011
...	... ...
...	... ...
9	011 1001
+	010 1011
–	010 1101
=	011 1101
A	100 0001
B	100 0010
...	... ...
O	100 1111
P	101 0000
Q	101 0001
...	... ...
...	... ...
Z	101 1010

ASCII code

Activity 14

How many possible codewords are there, using the ASCII system?

This code is not particularly efficient and for computers with limited memory space (e.g. hand-held calculators) often different codes are used.

One particular code in which the use of particular letters or numbers is very varied is called a **Huffman Code**.

As an example, consider a code needed for just five letters, say,

E   A   M   N   T

in which they are listed in order of decreasing frequency; that is, E is used more than A, A more than M, etc.

A possible Huffman code for five letters is shown below. The code for each letter is found by using a '1' for a left hand branch, and '0' for a right hand branch. So E is coded as '1', A as '0 0', etc., as shown.

Letter	Code
E	1
A	0 0
M	0 1 0
N	0 1 1 0
T	0 1 1 1

Why is this an efficient way of coding for this problem?

Note that there is no need to put gaps between codes for different letters as there can be no confusion, as you will see in the next example.

Example

Decode   0 1 1 0 0 0 0 1 0 1 0 1 0 0 0 0 1 1 0.

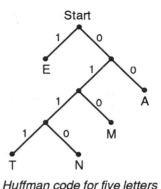

Huffman code for five letters

Solution

Using the diagram (or the table) you can follow through the code, stopping when any letter is reached:

0 1 1 0 : 0 0 : 0 1 0 : 1 : 0 1 0 : 0 0 : 0 1 1 0

N : A : M : E : M : A : N

Exercise 8B

1. Decode the following, using the Huffman code above

   (a) 0 1 0 0 0 0 1 1 0 0 0 0 1 1 1 1 0 1 1 1 1 0 0

   (b) 0 1 0 1 1 0 1 1 1 0 1 1 1 1 0 0 0 1 0 0 0 0 1 1
      1 0 1 1 1 1 0 1 1 0

2. Design a Huffman code if the only codewords used are as shown below and all words are used equally frequently.

BUS	CUPS	MUSH	PUSS
SIP	PUSH	CUSS	HIP
PUP	PUPS	HIPS	

8.7   Miscellaneous Exercises

1. The list below shows the International Morse Code (in 'dots and dashes') for some letters of the alphabet.

   J •――― K ―•― L •―•• M ―― N ―•

   Using these letters only,

   (a) give an example to show that in Morse Code even a single error can go undetected;

   (b) give an example of 7 dots and dashes to show that, unless a pause is left between letters, a message received in Morse Code may be decoded in more than one way.

2. A new furniture mail order company is designing a coding system for its variety of products. Information required to be coded includes:

   type of product
   size
   colour
   price
   catalogue number.

   Design a bar code system for identifying products in this company. Explain the rationale behind your design.

3. Research into one of the commonly used codes not covered earlier, and write a report outlining

   (a) the design of the code used

   (b) how it works in practice

   (c) advantages and disadvantages of the code.

4. Design a new coding system to solve a particular practical problem.

9 THEORY OF CODES

After studying this chapter you should

- understand what is meant by noise, error detection and correction;
- be able to find and use the Hamming distance for a code;
- appreciate the efficiency of codes;
- understand what is meant by a linear code and parity-check matrix;
- be able to decode a transmitted word using the parity check matrix.

9.0   Introduction

In this chapter you will look in some depth at the way in which codes of varying construction can be used both to detect and sometimes correct errors. You have already seen in Chapter 8 how check digits for ISBN numbers and bar codes are used to detect errors. In this chapter you will look at codes relevant to data transmission, for example the transmission of pictures from Mars to the Earth, and see how such codes are designed.

9.1   Noise

To take an example, in TV broadcasting the message for transmission is a picture in the studio. The camera converts this into a 625-row array of packages of information, each package denoting a particular colour. This array, in the form of an electrical signal, is broadcast via antennae and the atmosphere, and is finally interpreted by the receiving set in the living room. The picture seen there differs somewhat from the original, errors having corrupted the information at various stages in the channel of communication. These errors may result in effects varying from subtle changes of colour tone to what looks like a violent snowstorm. Technically, the errors are all classified as **noise**.

What form does 'noise' take in telephone calls?

A model of data transmission is shown below.

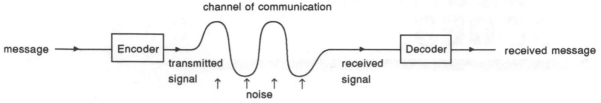

Normally, the message is encoded, the signal transmitted to the receiver, and then decoded with a received message. It is in the transmission that noise can affect the signal.

For example, the Mariner 9 spacecraft in 1971 sent television pictures of the planet Mars across a distance of 84 million miles. Despite a very low power transmitter, the space-probe managed to send data which eventually resulted in very high quality pictures being shown on our screens. This was in part largely due to the sophisticated coding system used.

As a very simple example, consider a code which has four **codewords**:

$$C = \big\{ (00),\ (01), (10), (11) \big\}$$

Each codeword has **length** 2, and all digits are either 0 or 1. Such codes are called **Binary Codes**.

Could you detect an error in the transmission of any of these codewords?

One way to detect an error, would be to repeat each codeword, giving a new code

$$C_1 = \big\{ (0000),\ (0101), (1010), (1111) \big\}$$

Here each pair of digits is repeated.

Can Code C_1 detect a single error?

For example, if the codeword $(0\ 1\ 0\ 1)$ was corrupted to $(1\ 1\ 0\ 1)$ it is clear that an error can be detected, as $(1\ 1\ 0\ 1)$ is **not** one of the codewords.

Can a single error in a codeword be corrected?

This is not as straightforward to answer since, for example, $(1\ 1\ 0\ 1)$ could have also been $(1\ 1\ 1\ 1)$ with one error, as well as $(0\ 1\ 0\ 1)$. So this code can detect a single error but cannot correct it. It should also be added that the **efficiency** (or rate) of this code is given by

$$\frac{\text{number of original message bits}}{\text{length of codeword}} = \frac{2}{4} = \frac{1}{2},$$

since each codeword in the original message had only two digits (called **bits**).

Activity 1

Consider a code designed to specify one of four possible directions

up	down	left	right
(0 0 0)	(1 1 0)	(0 1 1)	(1 0 1)

Can this code detect any single error made during the transmission of a codeword? Can it correct it?

Often codes include a **parity check** so that, for example, the code C is transformed to C_1 as shown below.

C	C_1
0 0	0 0 0
0 1	0 1 1
1 0	1 0 1
1 1	1 1 0

The extra last digit in C_1 is 0 if the sum of the digits modulo 2 is zero if even, or 1 if odd. (Modulo 2 means $0+0=0$, $0+1=1$, but $1+1=0$, etc.)

Can Code C_1 detect errors now?

Using the previous definition, the efficiency of Code C_1 is $\frac{2}{3}$.

None of the codes considered so far can correct errors.

Activity 2

Design a code containing 4 codewords, each of length 6, which can detect and correct a single error.

9.2 Error correction

Clearly codes which can both detect and correct errors are of far greater use - but the efficiency will decrease, since extra essentially redundant information will have to be transmitted.

For example, here is a code that can be used to idenfy four directions:

up	down	left	right
(0 0 0 0 0 0)	(1 1 1 0 0 0)	(0 0 1 1 1 0)	(1 1 0 0 1 1)

The length of each codeword is 6, but since the number of message bits is essentially 2, i.e. the code could consist of

$$(0\ 0),\ (1\ 1),\ (0\ 1),\ (1\ 0)$$

and its efficiency is $\dfrac{2}{6} = \dfrac{1}{3}$. But, as you see, it can **correct** single errors.

Activity 3

The following codewords from the above code have been received. Assuming that only one error has been made in the transmission of each codeword, determine if possible the actual codeword transmitted:

(a) (1 0 0 0 0 0)   (b) (1 1 0 0 0 0)   (c) (0 1 0 0 1 1)

Can the code above detect if 2 errors have been made in the transmission of a codeword?

Activity 4   Codes

Consider Code 5 given in Appendix 5. Find out how many errors this code can detect and correct by considering, for example, codewords such as

(a) (1 1 0 0 0 0 0)   (b) (0 1 1 1 1 1 1)   (c) (1 0 0 0 1 0 0)

which are in error.

By now you should be beginning to get a feel for what is the important characteristic of a code for the determination of the number of errors that can be detected and corrected. The crucial concept is that of **distance** between codewords.

The **distance** between any two codewords in a code is defined as the sum of the number of actual differences between the codewords; for example

$$d\big((111),\ (010)\big)\ \ = 2,$$

since the first and third digit are different;

whilst $d\big((0101),\ (1011)\big) = 3.$

The **Hamming distance** is defined as the **minimum** distance between any two codewords in the code and is usually denoted by δ.

Example

Determine the Hamming distance for the code with codewords

$$(1\ 1\ 0\ 0\ 0),\ (0\ 0\ 1\ 0\ 1),\ (1\ 0\ 1\ 0\ 1),\ (1\ 1\ 1\ 1\ 1)$$

Solution

You must first find distances between all the codewords.

$$d\left((11000),(00101)\right) = 4$$

$$d\left((11000),(10101)\right) = 3$$

$$d\left((11000),(11111)\right) = 3$$

$$d\left((00101),(10101)\right) = 1 \quad \leftarrow \text{Hamming distance } \delta = 1$$
$$\text{(minimum of 1, 2, 3 and 4)}$$

$$d\left((00101),(11111)\right) = 3$$

$$d\left((10101),(11111)\right) = 2$$

Why is the Hamming distance crucial for error detection and correction?

Activity 5   Hamming distance

Determine the Hamming distance for

Codes 1, 2, 3, 4 and **5**

given in Appendix 5.

To try and see the connection between the Hamming distance, δ, and the number of errors that can be detected or corrected, you will consider Codes 1 to 5 from Appendix 5.

Activity 6

Copy and complete this table.

Code	Hamming distance	Errors corrected	Errors detected
1	2	0	1
2	3	1	1
3	...	...	...
4	...	...	...
5	...	...	...

Also add on to the table any other codes considered so far. Can you see a pattern?

The first thing that you probably noticed about the data in the table for Activity 6 is that the results are different depending on whether n, the number of bits, is even or odd. It looks as if, for

$\delta = 2$, you can detect 1 error but correct 0 errors

$\delta = 3$, you can detect 1 error and correct 1 error.

How do you think the pattern continues for $\delta = 4$ and 5?

Activity 7

Construct a code for which $\delta = 4$. How many errors can it detect or correct? Similarly, construct a code for which $\delta = 5$ and again determine how many errors it can detect or correct. Can you suggest a generalisation of the results?

As can be seen from Activities 6 and 7, there is a distinct pattern emerging. For

δ odd,   the code can correct and detect up to $\frac{1}{2}(\delta - 1)$ errors.

δ even,   the code can correct up to $\frac{1}{2}(\delta - 2)$ errors, and detect up to $\frac{1}{2}\delta$ errors.

Activity 8   Validating the results

Check that the above result holds for Code 9 in Appendix 5.

Exercise 9A

1. The '2 out of 5' code consists of all possible words of length 5 which have exactly two 1 s; for example, (1 0 1 0 0) belongs to the code but (1 1 0 1 0) does not.

   List all possible codewords and explain why this code is particularly useful for the transmission of numeric data. What is the Hamming distance for this code?

2. Analyse the '3 out of 7' code, defined in a similar way to the '2 out of 5' code in Question 1. Determine its Hamming distance and hence find out how many words it can detect and correct.

3. Determine the Hamming distance for Code 7 in Appendix 5. Hence find out how many errors this code can detect and correct.

4. Show that the code

   $$C_4 = \{(00000), (11000), (00011), (11111)\}$$

   can detect but not correct single errors in transmission.

9.3 Parity check matrix

The main challenge of coding theory is to find good effective codes - that is, ones which transmit information efficiently yet are able to detect and correct a suitable number of errors.

Remember that the length of a code is the number of bits of its codewords - this is usually referred to as length n.

The codes used in the previous unit and those constructed here add another $(n - k)$ check bits to each message of length k to make a codeword of length n. For example, Code 3

$$0\,0\,0\,0$$
$$0\,1\,0\,1$$
$$1\,0\,1\,0$$
$$1\,1\,1\,1$$

is found by adding two check digits to each codeword of length 2, namely

$$0\,0$$
$$0\,1$$
$$1\,0$$
$$1\,1$$

So here $k = 2$, $n = 4$, and there are $n - k = 4 - 2 = 2$ check digits. The number k is called the **dimension** of the code and, as you saw in Section 9.1, the **efficiency** (or rate) is given by $\dfrac{k}{n}$.

A code of length n, with k message bits, is called an (n, k) code.

Example

Show that Code 4 is a $(4, 2)$ code.

Solution

Since the codewords of Code 4 are

$$0\ 0\ 0\ 0$$
$$1\ 1\ 0\ 0$$
$$0\ 0\ 1\ 1$$
$$1\ 1\ 1\ 1$$

then $n = 4$, and $k = 2$ since two columns are repeated. (Alternatively you might like to think of it in terms of 4 codewords, which could be coded using codewords of length 2; e.g. $(0\ 0)$, $(1\ 1)$, $(1\ 0)$, $(0\ 1)$; thus $k = 2$ and two more bits have been added to give $n = 4$.)

Thus Code 4 is a $(4, 2)$ code.

A vector and matrix notation will be adopted, writing a codeword $\mathbf{x}$ as a row vector; for example $[1\ 1\ 0\ 1]$.

The transpose of $\mathbf{x}$ is a column vector, $\mathbf{x'} = \begin{bmatrix} 1 \\ 1 \\ 0 \\ 1 \end{bmatrix}$

Let $\mathbf{x}$ be a codeword with n bits, so that

$$\mathbf{x} = \begin{bmatrix} x_1 & x_2 & \dots & x_n \end{bmatrix} \qquad (1 \times n \text{ matrix})$$

and

$$\mathbf{x'} = \begin{bmatrix} x_1 \\ x_2 \\ \dots \\ x_n \end{bmatrix}$$

For an (n, k) code, a **parity check matrix**, H, is defined as an $(n - k) \times n$ matrix such that

$$\mathbf{H}\mathbf{x'} = 0$$

and when no row of H consists just of zeros.

An alternative way of finding k is to write (when possible) the number of codewords in the code as a power of 2. This power is k, the dimension;

$$\text{i.e. no. of codewords} = 2^k.$$

Check this result for Code 3 and Code 4.

Example

Show that $\quad H = \begin{bmatrix} 1 & 1 & 0 & 0 \\ 1 & 1 & 1 & 1 \end{bmatrix}$

is a parity check matrix for Code 4.

Solution

For Code 4, $\mathbf{x}_1 = [0\ 0\ 0\ 0]$, $\mathbf{x}_2 = [1\ 1\ 0\ 0]$, $\mathbf{x}_3 = [0\ 0\ 1\ 1]$,

and $\quad H\mathbf{x}_1' = \begin{bmatrix} 1 & 1 & 0 & 0 \\ 1 & 1 & 1 & 1 \end{bmatrix}\begin{bmatrix} 0 \\ 0 \\ 0 \\ 0 \end{bmatrix} = \begin{bmatrix} 0 \\ 0 \end{bmatrix}$

$$H\mathbf{x}_2' = \begin{bmatrix} 1 & 1 & 0 & 0 \\ 1 & 1 & 1 & 1 \end{bmatrix}\begin{bmatrix} 1 \\ 1 \\ 0 \\ 0 \end{bmatrix} = \begin{bmatrix} 1+1 \\ 1+1 \end{bmatrix} = \begin{bmatrix} 0 \\ 0 \end{bmatrix}$$

(arithmetic is modulo 2)

$$H\mathbf{x}_3' = \begin{bmatrix} 1 & 1 & 0 & 0 \\ 1 & 1 & 1 & 1 \end{bmatrix}\begin{bmatrix} 0 \\ 0 \\ 1 \\ 1 \end{bmatrix} = \begin{bmatrix} 0 \\ 1+1 \end{bmatrix} = \begin{bmatrix} 0 \\ 0 \end{bmatrix}$$

$$H\mathbf{x}_4' = \begin{bmatrix} 1 & 1 & 0 & 0 \\ 1 & 1 & 1 & 1 \end{bmatrix}\begin{bmatrix} 1 \\ 1 \\ 1 \\ 1 \end{bmatrix} = \begin{bmatrix} 1+1 \\ 1+1+1+1 \end{bmatrix} = \begin{bmatrix} 0 \\ 0 \end{bmatrix}$$

Hence H is a parity check matrix for Code 4.

In fact, the codewords $\mathbf{x} = (x_1\ x_2\ x_3\ x_4)$ of Code 4 are precisely the solutions of

$$\begin{bmatrix} 1 & 1 & 0 & 0 \\ 1 & 1 & 1 & 1 \end{bmatrix}\begin{bmatrix} x_1 \\ x_2 \\ x_3 \\ x_4 \end{bmatrix} = \begin{bmatrix} 0 \\ 0 \end{bmatrix}$$

$$\Rightarrow \quad \begin{cases} x_1 + x_2 = 0 & \text{(modulo 2)} \\ x_1 + x_2 + x_3 + x_4 = 0 & \text{(modulo 2)} \end{cases}$$

Activity 9

Find all solutions, modulo 2, of the two equations above.

This property, that $H\mathbf{x}' = 0$ has as its solution the codewords of the code, leads us into a method of finding a parity check matrix. Consider for example Code 2 from Appendix 5. This has length 6 and all codewords in Code 2 satisfy

$$x_2 + x_3 + x_4 = 0 \quad \text{(modulo 2)}$$
$$x_1 + x_3 + x_5 = 0 \quad \text{(modulo 2)}$$
$$x_1 + x_2 + x_6 = 0 \quad \text{(modulo 2)}.$$

In matrix form these can be written as

$$\begin{bmatrix} 0 & 1 & 1 & 1 & 0 & 0 \\ 1 & 0 & 1 & 0 & 1 & 0 \\ 1 & 1 & 0 & 0 & 0 & 1 \end{bmatrix} \begin{bmatrix} x_1 \\ x_2 \\ x_3 \\ x_4 \\ x_5 \\ x_6 \end{bmatrix} = \begin{bmatrix} 0 \\ 0 \\ 0 \end{bmatrix}$$

Since $n = 6$, and $k = 3$, the above 3×6 matrix H is a parity check matrix for Code 2.

Example

Find a parity check matrix for Code 3 from Appendix 5.

Solution

For Code 3, $n = 4$ and $k = 2$, so it is a (4, 2) Code and H will be a 2×4 matrix. Now for all codewords in Code 3,

$$x_1 + x_3 = 0 \quad \text{(modulo 2)}$$
$$x_2 + x_4 = 0 \quad \text{(modulo 2)}$$
$$x_1 + x_2 + x_3 + x_4 = 0 \quad \text{(modulo 2)}.$$

Only two of these equations are needed, so, for example, a possible parity check matrix is given by

$$\mathbf{H} = \begin{bmatrix} 1 & 0 & 1 & 0 \\ 0 & 1 & 0 & 1 \end{bmatrix}.$$

Activity 10

Find two other possible parity check matrices for Code 3.

Activity 11

Find a parity check matrix for Code 5 in Appendix 5.

9.4 Decoding using parity check matrices

Returning to the vector notation introduced earlier, suppose that the codeword **x** is transmitted, resulting in word

$$\mathbf{r} = \mathbf{x} + \mathbf{e}$$

being received. Hence **e** is the error word that has corrupted **x**.

For example, suppose the codeword transmitted is $(1\ 1\ 1\ 0\ 0)$ but that the received word is $\mathbf{r} = (0\ 1\ 1\ 0\ 0)$. This means that the error word is given by

$$\mathbf{e} = (1\ 0\ 0\ 0\ 0).$$

Parity check matrices can be very useful for finding out the most likely errors in transmission in the case of **linear** codes.

A linear code, with parity check matrix H consists of all the words **x** which satisfy the equation

$$H\mathbf{x'} = 0.$$

Example

Show that Code 3 is linear.

Solution

A parity check matrix for Code 3 is given by

$$H = \begin{bmatrix} 1 & 0 & 1 & 0 \\ 0 & 1 & 0 & 1 \end{bmatrix} \text{ (see Activity 10)}$$

Now the equation $H\mathbf{x'} = \mathbf{0}$ can be written as

$$\begin{bmatrix} 1 & 0 & 1 & 0 \\ 0 & 1 & 0 & 1 \end{bmatrix} \begin{bmatrix} x_1 \\ x_2 \\ x_3 \\ x_4 \end{bmatrix} = \begin{bmatrix} 0 \\ 0 \end{bmatrix}$$

or
$$x_1 + x_3 = 0$$
$$x_2 + x_4 = 0.$$

If $x_1 = 1$, then $x_3 = 1$ (remember addition is modulo 2) whereas $x_1 = 0$ means $x_3 = 0$. Similarly for x_2 and x_4. This gives the following codewords

$$1\ 0\ 1\ 0$$

$$1\ 1\ 1\ 1$$

$$0\ 1\ 0\ 1$$

$$0\ 0\ 0\ 0$$

which is Code 3.

The importance of linear codes is that if $\mathbf{x}$ and $\mathbf{y}$ are two codewords in the linear code with parity check matrix H,

$$H(\mathbf{x}+\mathbf{y})' = H(\mathbf{x'}+\mathbf{y'})$$

$$= H\mathbf{x'}+H\mathbf{y'}$$

$$= \mathbf{0}+\mathbf{0}$$

$$= \mathbf{0}.$$

Hence $\mathbf{x}+\mathbf{y}$ is also a codeword.

The reverse is also true. That is, for every possible codeword $\mathbf{x}$ and $\mathbf{y}$, if $\mathbf{x}+\mathbf{y}$ is also a codeword, then the code is linear.

Activity 12

Show that the code with codewords

$$0\ 0\ 0\ 0\ 0\ 0\ 0\ 0\ 0\ 0$$

$$0\ 1\ 1\ 0\ 1\ 1\ 0\ 0\ 1\ 0$$

$$1\ 0\ 0\ 1\ 0\ 0\ 1\ 1\ 0\ 1$$

$$1\ 1\ 1\ 1\ 1\ 1\ 1\ 1\ 1\ 1$$

is a linear code.

Now for a linear code, if $\mathbf{r}=\mathbf{x}+\mathbf{e}$ is the received word, then

$$H\mathbf{r'}= H(\mathbf{x}+\mathbf{e})'$$

$$= H(\mathbf{x'}+\mathbf{e'})$$

$$= H\mathbf{x'}+H\mathbf{e'}$$

$$= \mathbf{0}+H\mathbf{e'}$$

$$\Rightarrow \quad \boxed{H\mathbf{r'}= H\mathbf{e'}}$$

Note that this result shows that $H\mathbf{r'}$ is independent of the codeword transmitted.

Example

Suppose a codeword from Code 2 is received as $\mathbf{r} = (1\,1\,0\,1\,1\,1)$. What is the most likely codeword sent?

Solution

$$\mathbf{H\,r} = \begin{bmatrix} 0 & 1 & 1 & 1 & 0 & 0 \\ 1 & 0 & 1 & 0 & 1 & 0 \\ 1 & 1 & 0 & 0 & 0 & 1 \end{bmatrix} \begin{bmatrix} 1 \\ 1 \\ 0 \\ 1 \\ 1 \\ 1 \end{bmatrix} = \begin{bmatrix} 0 \\ 0 \\ 1 \end{bmatrix}$$

This result corresponds to the **last** column of H, so you would conclude that the codeword sent is in error in its last digit, and should have been

$$1\,1\,0\,1\,1\,0.$$

Activity 13

For the following words

(a) $1\,1\,1\,0\,0\,0\,0$ (b) $0\,1\,1\,1\,0\,1\,1$

use a parity check matrix to determine which codewords from Code 5 were actually transmitted.

So the parity check matrix provides a means of finding the most likely error in transmission for linear codes.

9.5   Cyclic codes

Many codes have been designed to meet a variety of situations. One special sort of code is called a **cyclic** code.

Codes are called **cyclic** if they have the property that whenever

$$\mathbf{x} = (x_1\ x_2\ \dots\ x_n)$$

is a codeword, then so is $\mathbf{x}^*$ defined by

$$\mathbf{x}^* = (x_n\ x_1\ x_2\ \dots\ x_{n-2}\ x_{n-1}).$$

Example

Show that Code 1 in Appendix 5 is a cyclic code.

Solution

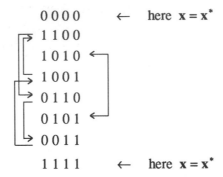

$$
\begin{array}{ll}
0\,0\,0\,0 & \leftarrow \quad \text{here } \mathbf{x} = \mathbf{x}^* \\
1\,1\,0\,0 & \\
1\,0\,1\,0 & \\
1\,0\,0\,1 & \\
0\,1\,1\,0 & \\
0\,1\,0\,1 & \\
0\,0\,1\,1 & \\
1\,1\,1\,1 & \leftarrow \quad \text{here } \mathbf{x} = \mathbf{x}^*
\end{array}
$$

Hence the code is cyclic.

Activity 14

Show that Code 3 is a cyclic code.

Exercise 9B

1. Consider the linear code whose eight codewords are as follows:

0 0 1 1 1 0 1	0 1 0 1 0 1 1
0 1 1 0 1 1 0	1 0 0 0 1 1 1
1 0 1 1 0 1 0	1 1 0 1 1 0 0
1 1 1 0 0 0 1	0 0 0 0 0 0 0.

   (a) Find the distance between any two codewords, and hence find the minimum distance.

   (b) Find the number of errors in a transmitted codeword which can be simultaneously detected and corrected by this code.

   (c) A codeword is transmitted and the binary word 1 0 0 1 1 0 0 is received. Which codeword is most likely to have been transmitted?

2. Let C be the code with the parity check matrix

   $$H = \begin{bmatrix} 1 & 1 & 0 & 0 \\ 0 & 0 & 1 & 1 \end{bmatrix}.$$

   Find the codewords of C and write down the minimum distance of C.

3. The eight codewords of a linear code are as follows:

   0 0 0 0 0 0 0
   0 0 1 1 1 0 1
   1 0 0 1 0 1 1
   1 0 1 0 1 1 0
   1 1 0 0 1 0 1
   1 1 1 1 0 0 0
   0 1 1 0 0 1 1
   0 1 0 1 1 1 0

   (a) State the minimum distance of this code.

   (b) How many errors per received codeword can this code

   (i) correct   (ii) detect?

   (c) A codeword is transmitted and the binary word 0 1 0 0 1 0 1 is received. Which of the eight codewords is most likely to have been the one transmitted?

4. Show that the code

0 0 0 0 0 0	1 0 1 0 1 1
0 0 0 1 1 1	1 0 1 1 0 0
0 1 1 0 0 1	1 1 0 0 1 0
0 1 1 1 1 0	1 1 0 1 0 1

   is linear and find a parity check matrix. Use it to decode the received message 0 1 0 1 1 0.

9.6 Miscellaneous Exercises

1. The code C has parity-check matrix

$$H = \begin{bmatrix} 0 & 0 & 0 & 1 & 1 & 1 & 1 \\ 0 & 1 & 1 & 0 & 0 & 1 & 1 \\ 1 & 0 & 1 & 0 & 1 & 0 & 1 \end{bmatrix}$$

(a) Write down the length, the dimension, and the rate of this code.

(b) A codeword of C is transmitted and incorrectly received as 0111000. Find the possible error and the transmitted codeword, assuming that only one error has occurred.

2. Consider the code whose codewords are 0000000000, 0110110010, 1001001101, 1111111111.

(a) How many errors does this code **simultaneously** correct and detect?

(b) If a message is received as 0110111111, which codeword is most likely to have been transmitted?

(c) Is this code a linear code?

(In each part, give reasons for your answer.)

3. One of the codewords of a cyclic code is 1001110.

(a) List the other six words of the code.

(b) What is the Hamming distance of this code?

(c) How many errors in a codeword can be simultaneously detected and corrected? Give a brief reason for your answer.

(d) Show that the code is not linear. What is the minimum number of codewords which need to be added to make this code linear?

(e) The matrix

$$\begin{bmatrix} 1 & 1 & 1 & 0 & 1 & 0 & 0 \\ 0 & 1 & 1 & 1 & 0 & 1 & 0 \\ 1 & 1 & 0 & 1 & 0 & 0 & 1 \end{bmatrix}$$

is a parity check matrix for this code. Show how to use it to correct the received message

0 1 1 0 1 1 1 1 1 1 0 1 0 0 0 0 1 1 1 1 1.

*4. Consider the linear code C whose eight codewords are as follows:

0000000000, 1001011100,
0100101110, 1101110010,
0010010111, 1011001011,
0110111001, 1111100101.

(a) What is the minimum distance of this code? How many errors in a transmitted codeword can this code simultaneously correct and detect?

(b) The matrix

$$H = \begin{bmatrix} 1 & 0 & 0 & 1 & 0 & 0 & 0 & 0 & 0 & 0 \\ 0 & 1 & 0 & 0 & 1 & 0 & 0 & 0 & 0 & 0 \\ 1 & 0 & 1 & 0 & 0 & 1 & 0 & 0 & 0 & 0 \\ 1 & 1 & 0 & 0 & 0 & 0 & 1 & 0 & 0 & 0 \\ 1 & 1 & 1 & 0 & 0 & 0 & 0 & 1 & 0 & 0 \\ 0 & 1 & 1 & 0 & 0 & 0 & 0 & 0 & 1 & 0 \\ 0 & 0 & 1 & 0 & 0 & 0 & 0 & 0 & 0 & 1 \end{bmatrix}$$

is a parity-check matrix for C. Use it to decode the received word 0010001111.

10 LOGIC

Objectives

After studying this chapter you should

- understand the nature of propositional logic;
- understand the connectives NOT, OR, AND;
- understand implication and equivalence;
- be able to use truth tables;
- be able to identify tautology and contradiction;
- be able to test the validity of an argument.

10.0 Introduction

It may seem unusual for philosophical ideas of logic based on intuition to be represented mathematically, however, the mathematics that has developed to describe logic has, in recent years, been crucial in the design of computer circuits and in automation.

Charles L Dodgson (1832 -1898) who under the pseudonym *Lewis Carroll* wrote 'Alice in Wonderland', was an Oxford mathematician who wrote about logic. One example of his logic problems concerns Mrs Bond's ducks.

Activity 1   Do ducks wear collars?

The following lines are taken from Lewis Carroll's book 'Symbolic Logic' first published in 1897.

> "All ducks in this village, that are branded 'B' belong to Mrs Bond;
>
> Ducks in this village never wear lace collars, unless they are branded 'B';
>
> Mrs Bond has no grey ducks in this village."

Is the conclusion 'no grey ducks in this village wear lace collars' valid?

10.1 The nature of logic

The Greek philosopher *Aristotle* (384-322 BC) is considered to be
the first to have studied logic in that he formed a way of
representing **logical propositions** leading to a conclusion.
Aristotle's theory of **syllogisms** provides a way of analysing
propositions given in the form of statements.

For example, here are some propositions.

>All apples are fruits.

>No toothache is pleasant.

>Some children like chocolate.

>Some cheese is not pasteurised.

In each of these statements a **subject**, S (e.g. apples) is linked to a
predicate, P (e.g. fruits). The **quantity** of each subject is indicated
by the word 'all', 'no' or 'some'.

Statements can also be described as **universal** ('all' or 'no') or
particular ('some') and **affirmative** or **negative**.

The four statements above can be described in this way:

>all S is P   universal affirmative

>no S is P   universal negative

>some S is P   particular affirmative

>some S is not P   particular negative .

Aristotle described an argument by linking together three
statements; two statements, called **premises,** lead to a third
statement which is the **conclusion** based on the premises. This
way of representing an **argument** is called a **syllogism.**

Example

>If fruits are tasty

>and apples are fruits
>_____

>then apples are tasty.

In this example of a syllogism the conclusion of the argument has
apples as subject (S) and tasty as predicate (P). The first premise
includes P and the second premise includes S and both the
premises include 'fruit', which is known as the middle term (M).

The syllogism can therefore be described as:

> If M is P
>
> and S is M
> _____
> then S is P.

Activity 2   Finding the figure

By removing all the words the last example can be described as

> M P
>
> S M
> ___
> S P

On the assumption that

- the conclusion of the argument must be SP,
- the first premise must contain P,
- the second premise must contain S,
- both the first and the second premise must contain M,

find the three other arrangements.

Together these are known as the **four figures** of the syllogism.

Not all syllogisms are valid

Each of the four figures can be universal or particular, affirmative or negative, but not all these arrangements give valid arguments.

Example

Is this valid?

> No M is P
>
> All S is M
> _____
> Some S is P.

Solution

An example of this syllogism might be:

No animal with 4 legs is a bird

All cats are animals with 4 legs
_____

Some cats are birds.

Obviously this arrangement is an invalid argument!

Activity 3   Valid or invalid?

Decide if these syllogisms are valid.

	(a)	Some M is P	(b)	All P is M

(a) Some M is P (b) All   P is M

  All   S is M   No   S is M

  All   S is P   No   S is P

(c) No   M is P (d) Some P is M

  All   M is S   All   M is S

  Some S is P   Some S is P

10.2 Combining propositions

Modern logic is often called **propositional logic**; the word 'proposition' is defined as a statement that is either **true** or **false**. So far, a variety of propositions have been considered, such as premises and conclusions to an argument.

For example, consider the statement

'the water is deep'.

It is not possible to say if this is true or false unless the word 'deep' is defined and, without a precise definition, this cannot be called a proposition.

Example

- **p** stands for the proposition 'January has 31 days', which is true.

- **q** stands for the proposition '$4 + 7 = 10$', which is false.

- 'What a hot day' is not a proposition because it is not in subject-predicate form; also the word 'hot' is not defined.

Negation NOT ~

Each proposition has a corresponding negation and, if the proposition is denoted by **p**, the negation of the proposition is denoted by ~**p**, read as 'not **p**'.

Example

If **p** is the proposition 'the table is made of pine',

then   ~**p** is the proposition 'the table is not made of pine'.

If **q** is the proposition 'the sack is empty', then ~ **q** is the statement 'the sack is not empty'. It is not correct to assume that the negation is 'the sack is full', since the statement 'the sack is not empty' could mean 'the sack is only partly full'.

Connectives

Simple propositions such as

'Elgar composed the Enigma Variations'

'Elgar lived in Malvern'

can be joined by the **connective** 'and' to form a **compound proposition** such as

'Elgar composed the Enigma Variations **and** lived in Malvern.'

A **compound proposition** can be described as a proposition made up of two or more simple propositions joined by connectives.

There is a variety of connectives which will now be defined.

Conjunction AND ∧

If two propositions are joined by the word AND to form a compound statement, this is called a **conjunction** and is denoted by the symbol ∧.

Example

If **p** is the proposition 'the sun is shining'

and **q** is the proposition 'Jack is wearing sunglasses',

then **p** ∧**q** represents the conjunction 'the sun is shining AND Jack is wearing sunglasses'

Disjunction OR ∨

If two statements are joined by the word OR to form a compound proposition, this is called a **disjunction** and is denoted by the symbol ∨.

Example

If **p** is the proposition 'Ann is studying geography'

and **q** is the proposition 'Ann is studying French'

then the disjunction **p** ∨**q** is the compound statement

 'Ann is studying geography OR French.'

The word 'OR' in this context can have two possible meanings.

Can Ann study both subjects?

Think about the meaning of these two sentences.

> 'I can deliver your coal on Wednesday or Thursday.'

> 'My fire can burn logs or coal.'

The first sentence implies that there is only one delivery of coal and illustrates the **exclusive** use of OR, meaning 'or' but not 'both'. The coal can be delivered on Wednesday or Thursday, but would not be delivered on both days.

The second sentence illustrates the **inclusive** use of 'OR' meaning that the fire can burn either logs or coal, or both logs and coal.

The word 'OR' and the symbol '∨' are used for the **inclusive** OR, which stands for 'and/or'.

The exclusive OR is represented by the symbol ⊕.

Activity 4   Exclusive or inclusive?

Write down three English sentences which use the inclusive OR and three which use the exclusive OR.

Now that a range of connectives is available propositions can be combined into a variety of compound propositions.

Example

Use **p**, **q** and **r** to represent affirmitive (or positive) statements and express the following proposition symbolically.

'Portfolios may include paintings or photographs but not collages.'

Solution

So, let **p** be 'portfolios may include paintings'

let **q** be 'portfolios may include photographs'

and let **r** be 'portfolios may include collages'.

The proposition therefore becomes

$$(p \lor q) \land \sim r.$$

Exercise 10A

1. For each of these compound propositions, use **p**, **q** and **r** to represent affirmative (or positive) statements and then express the proposition symbolically.

   (a) This mountain is high and I am out of breath.

   (b) It was neither wet nor warm yesterday.

   (c) During this school year Ann will study two or three subjects.

   (d) It is not true that $3+7=9$ and $4+4=8$.

2. Let **p** be 'the cooker is working', **q** 'the food supply is adequate' and **r** 'the visitors are hungry'. Write the following propositions in 'plain English':

   (a) $p \land \sim r$

   (b) $q \land r \land \sim p$

   (c) $r \lor \sim q$

   (d) $\sim r \lor (p \land q)$

   (e) $\sim q \land (\sim p \lor \sim r)$

10.3 Boolean expressions

The system of logic using expressions such as $p \lor q$ and $\sim p \land r$ was developed by the British mathematician *George Boole* (1815 - 1864).

The laws of reasoning were already well known in his time and Boole was concerned with expressing the laws in terms of a special algebra which makes use of what are known as **Boolean expressions**, such as $\sim a \land b.$

Activity 5   Using plain English

Define three propositions of your own, **p, q** and **r,** and write in plain English the meaning of these Boolean expressions.

1.	$q \wedge r$	3.	$\sim p \vee (q \wedge r)$
2.	$\sim p \wedge r$	4.	$r \wedge (\sim p \vee q)$

Using truth tables

In Section 10.2 a proposition was defined as a statement that is either true or false. In the context of logic, the integers 0 and 1 are used to represent these two states.

     0 represents false

     1 represents true.

Clearly, if a proposition **p** is true then ~**p** is false; also if **p** is false, then ~ **p** is true.   This can be shown in a **truth table**, as below.

p	~ p
0	1
1	0

The connectives, $\vee$ and $\wedge$ can also be defined by truth tables, as shown below.

p	q	p ∧ q
0	0	0
0	1	0
1	0	0
1	1	1

This truth table shows the truth values (0 or 1) of the conjunction **p ∧ q**.

Since **p ∧ q** means **p** AND **q,** then **p ∧ q** can only be true (ie 1) when **p** is true AND **q** is true.

If a negation is used, it is best to add the negation column, eg ~**p**, to the truth table.

Example

Construct the truth table for $p \wedge \sim q$.

Solution

p	q	$\sim q$	$p \wedge \sim q$
0	0	1	0
0	1	0	0
1	0	1	1
1	1	0	0

If **p** AND ~**q** are true (ie both are 1) then $p \wedge \sim q$ is true.

Exercise 10B

Construct truth tables for the following.

1. $q \vee r$
2. $\sim p \wedge r$
3. $p \vee \sim r$
4. $\sim p \vee \sim q$

10.4 Compound propositions

More complicated propositions can be represented by truth tables, building up parts of the expression.

Example

Construct the truth table for the compound proposition
$(a \vee b) \vee \sim c$.

Solution

a	b	c	$(a \vee b)$	$\sim c$	$(a \vee b) \vee \sim c$
0	0	0	0	1	1
0	0	1	0	0	0
0	1	0	1	1	1
0	1	1	1	0	1
1	0	0	1	1	1
1	0	1	1	0	1
1	1	0	1	1	1
1	1	1	1	0	1

Exercise 10C

Construct truth tables for the following:

1. $(a \vee b) \vee c$

2. $a \wedge (b \wedge c)$

3. $a \vee (b \vee c)$

4. $(a \wedge b) \wedge c$

5. $a \wedge (b \vee c)$

6. $(a \wedge b) \vee (a \wedge c)$

7. $a \vee (b \wedge c)$

8. $(a \vee b) \wedge (a \vee c)$

You should notice that some of your answers in this exercise are the same. What are the implications of this? Can you think of similar rules for numbers in ordinary algebra? What names are given to these properties?

10.5 What are the implications?

'If I win this race, then I will be in the finals.'

'If the light is red, then you must stop.'

These two sentences show another connective, 'if ... then ...' which is indicated by the symbol $\Rightarrow$.

$x \Rightarrow y$ is the compound proposition meaning that proposition x implies proposition y.

Returning to the compound proposition

'If I win this race, then I will be in the finals',

this can be written as $a \Rightarrow b$ where a is the proposition 'I win the race' and b is the proposition 'I will be in the finals'. The first proposition, a, 'I win this race', can be true or false. Likewise, the second proposition 'I will be in the finals' can be true or false.

If I win the race (a is true) and I am in the final (b is true) then the compound proposition is true ($a \Rightarrow b$ is true).

Activity 6   The implication truth table

If I fail to win the race (a is false) and I am not in the final (b is false), is the compound proposition $a \Rightarrow b$ true or false?

If I win the race but am not in the final (illness, injury), then is the compound proposition $a \Rightarrow b$ true or false?

By considering these two questions and two others, it is possible to build up a truth table for the proposition $a \Rightarrow b$. Think about the other two questions and their answers, and hence complete the following truth table.

a	b	a $\Rightarrow$ b
0	0	
0	1	
1	0	
1	1	

The values in this truth table often cause much argument, until it is realised that the connective $\Rightarrow$ is about implication and not about cause and effect.

It is not correct to assume that $a \Rightarrow b$ means a causes b or that b results from a.

In fact, the implication connective, $\Rightarrow$, is defined by the values shown in the truth table whatever the propositions that make up the compound proposition.

Consider the implication $a \Rightarrow b$

'If it is hot, it is June.'

The only way of being sure that this implication $a \Rightarrow b$ is false is by finding a time when it is hot but it isn't June; i.e. when a is true but b is false. Hence the truth table for $a \Rightarrow b$ is as follows:

a	b	a $\Rightarrow$ b
0	0	1
0	1	1
1	0	0
1	1	1

In logic, the two propositions which make up a compound proposition may not be related in the usual sense.

Example

'If Christmas is coming (C), today is Sunday (S).'

C	S	C $\Rightarrow$ S
0	0	1
0	1	1
1	0	0
1	1	1

However difficult it may seem to invent a meaning for this implication, the truth table will be exactly the same as before.

Exercise 10D

1. Give the truth values (1 or 0) of these propositions.

   (a) If all multiples of 9 are odd, then multiples of 3 are even.

   (b) If dogs have four legs then cats have four legs.

   (c) If the sea is blue, the sky is green.

   (d) Oxford is in Cornwall if Sheffield is in Yorkshire.

   (e) Pentagons have six sides implies that quadrilaterals have four sides.

2. If **a** represents 'the crops grow', **b** is 'I water the plants' and **c** is 'I spread manure', express these propositions in terms of **a**, **b** and **c**.

   (a) If I water the plants the crops grow.

   (b) I do not spread manure nor do I water the plants and the crops do not grow.

   (c) If I spread manure the crops grow.

   (d) The crops grow if I water the plants and do not spread manure.

   (e) If I do not water the plants, then I spread manure and the crops grow.

3. Using **a**, **b** and **c** from Question 2, interpret the following propositions.

   (a) $(a \wedge b) \vee (a \wedge c)$

   (b) $(c \vee \sim b) \Rightarrow \sim a$

   (c) $a \Rightarrow b \wedge c$

   (d) $\sim a \vee c \Rightarrow b$

10.6 Recognising equivalence

There is a difference between the proposition

'If it is dry, I will paint the door.'

and the proposition

'If, and only if, it is dry, I will paint the door.'

If **p** is 'it is dry' and **q** is 'I will paint the door', then $p \Rightarrow q$ represents the first proposition.

The second proposition uses the connective of **equivalence** meaning 'if and only if ' and is represented by the symbol ⇔, i.e. **p ⇔ q** represents the second proposition.

The truth table for **p ⇔ q** shown here is more obvious than the truth table for implication.

p	q	p ⇔ q
0	0	1
0	1	0
1	0	0
1	1	1

You will see that **p ⇔ q** simply means that the two propositions **p** and **q** are true or false together: this accounts for the use of the word 'equivalence'. Note that if you work out the truth table for **q ⇔ p** you will get the same results as for **p ⇔ q**.

Exercise 10E

1. If **a** is a true statement and **b** is false, write down the truth value of:

    (a) $a \Leftrightarrow \sim b$

    (b) $\sim b \Leftrightarrow \sim a$.

2. If **a** is 'the theme park has excellent rides', **b** is 'entrance charges are high' and **c** is 'attendances are large', write in plain English the meaning of:

    (a) $c \Leftrightarrow (a \wedge \sim b)$

    (b) $(\sim c \vee \sim b) \Rightarrow a$.

10.7 Tautologies and contradictions

In the field of logic, a **tautology** is defined as a compound proposition which is **always true** whatever the truth values of the constituent statements.

p	~ p	p∨ ~ p
0	1	1
1	0	1

This simple truth table shows that

   p∨ ~ p is a tautology.

The opposite of a tautology, called a **contradiction**, is defined as a compound proposition which is **always false** whatever the truth values of the constituent statements.

p	~p	p∧~p
0	1	0
1	0	0

This simple truth table shows that

$p \wedge \sim p$ is a contradiction.

Example

Is $\left[a \wedge (b \vee \sim b)\right] \Leftrightarrow a$ a tautology or a contradiction?

Solution

The clearest way to find the solution is to draw up a truth table. If the result is always true then the statement is a tautology; if always false then it is a contradiction.

a	b	~b	(b∨~b)	a∧(b∨~b)	[a∧(b∨~b)]⇔a
0	0	1	1	0	1
0	1	0	1	0	1
1	0	1	1	1	1
1	1	0	1	1	1

The truth table shows that, since the compound statement is always true, the example given is a tautology.

Exercise 10F

Decide whether each of the following is a tautology or a contradiction.

1. $(a \Rightarrow b) \Leftrightarrow (a \wedge \sim b)$
2. $\left[a \wedge (a \Rightarrow b)\right] \wedge \sim b$
3. $\sim(a \Rightarrow b) \Rightarrow \left[(b \vee c) \Rightarrow (a \vee c)\right]$

10.8 The validity of an argument

In Section 8.1 the idea of an argument was described as a set of premises (such as **p**, **q** and **r**) which leads to a conclusion (**c**):

$$
\begin{array}{c}
\mathbf{p} \\
\mathbf{q} \\
\mathbf{r} \\
\cdot \\
\cdot \\
\cdot \\
\hline
\mathbf{c}
\end{array}
$$

A **valid** argument is one in which, if the premises are true, the conclusion must be true. An **invalid** argument is one that is not valid. The validity of an argument can, in fact, be independent of the truth (or falsehood) of the premises. It is possible to have a valid argument with a false conclusion or an invalid argument with a true conclusion. An argument can be shown to be valid if $\mathbf{p} \wedge \mathbf{q} \wedge \mathbf{r} \wedge .... \Rightarrow \mathbf{c}$ is always true (i.e. a tautology).

Example

Represent the following argument symbolically and determine whether the argument is valid.

> If cats are green then I will eat my hat.
>
> I will eat my hat.
> _____
> Cats are green.

Solution

Write the argument as

$$
\begin{array}{c}
\mathbf{a} \Rightarrow \mathbf{b} \\
\mathbf{b} \\
\hline
\mathbf{a}
\end{array}
$$

The argument is valid if $(\mathbf{a} \Rightarrow \mathbf{b}) \wedge \mathbf{b} \Rightarrow \mathbf{a}$.

a	b	$(\mathbf{a} \Rightarrow \mathbf{b})$	$(\mathbf{a} \Rightarrow \mathbf{b}) \wedge \mathbf{b}$	$(\mathbf{a} \Rightarrow \mathbf{b}) \wedge \mathbf{b} \Rightarrow \mathbf{a}$
0	0	1	0	1
0	1	1	1	0
1	0	0	0	1
1	1	1	1	1

The truth table shows that the argument is not always true (i.e. it is not a tautology) and is therefore invalid. The second line in the truth table shows that the two premises $a \Rightarrow b$ and b can both be true with the conclusion a being false. In other words, the compound proposition $(a \Rightarrow b) \wedge b \Rightarrow a$ is not always true (i.e. it is not a tautology). Therefore the argument

$$a \Rightarrow b$$
$$\underline{b}$$
$$a$$

is invalid.

Exercise 10G

Determine whether these arguments are valid.

1.
$$a \Rightarrow b$$
$$\underline{a \Rightarrow c}$$
$$a \Rightarrow (b \wedge c)$$

2.
$$\sim b \Rightarrow \sim a$$
$$\underline{b}$$
$$a$$

3.
$$p \Rightarrow q$$
$$\underline{r \Rightarrow \sim q}$$
$$p \Rightarrow \sim r$$

4. Form a symbolic representation of the following argument and determine whether it is valid.

If I eat well then I get fat.

If I don't get rich then I don't get fat.

I get rich.

10.9 Miscellaneous Exercises

1. Denote the positive (affirmative) statements in the following propositions by a, b, c, ... and express each proposition symbolically.

   (a) Either you have understood this chapter, or you will not be able to do this question.

   (b) 64 and 169 are perfect squares.

   (c) $-4 > -9$ and $4 > -9$.

   (d) This is neither the right time nor the right place for an argument.

   (e) If the wind is blowing from the east, I will go sailing tomorrow.

   (f) The train standing at platform 5 will not leave unless all the doors are shut.

   (g) The telephone rang twice and there was no reply.

   (h) My friend will go to hospital if his back doesn't get better.

2. Draw up a truth table for these propositions:

   (a) $(p \vee \sim q) \Rightarrow q$

   (b) $[p \vee (\sim p \vee q)] \vee (\sim p \wedge \sim q)$

   (c) $(\sim p \vee \sim q) \Rightarrow (p \wedge \sim q)$

   (d) $\sim p \Leftrightarrow q$

   (e) $(\sim p \wedge q) \vee (r \wedge p)$

   (f) $(p \Leftrightarrow q) \Rightarrow (\sim p \wedge q)$

3. Decide whether each of the following is a tautology:

   (a) $\sim a \Rightarrow (a \Rightarrow b)$

   (b) $\sim (a \vee b) \wedge a$

   (c) $[a \wedge (a \Rightarrow b)] \Rightarrow a$

   (d) $(a \Rightarrow b) \Leftrightarrow \sim (a \wedge \sim b)$

4. Decide whether each of the following is a contradiction:

    (a) $(a \land b) \lor (\sim a \land \sim b)$

    (b) $(a \Rightarrow b) \Leftrightarrow (a \land \sim b)$

    (c) $\sim(a \land b) \lor (a \lor b)$

    (d) $(a \lor b) \Rightarrow \sim(b \lor c)$

5. Formulate these arguments symbolically using **p**, **q** and **r**, and decide whether each is valid.

    (a) If I work hard, then I earn money

    I work hard
    _____
    I earn money

    (b) If I work hard then I earn money

    If I don't earn money then I am not successful
    _____
    I earn money

    (c) I work hard if and only if I am successful

    I am successful
    _____
    I work hard.

    (d) If I work hard or I earn money then I am successful

    I am successful
    _____
    If I don't work hard then I earn money.

*6. *Lewis Carroll* gave many arguments in his book 'Symbolic Logic'. Decide whether the following arguments are valid.

    (a) No misers are unselfish.

    None but misers save egg-shells.
    _____
    No unselfish people save egg-shells.

    (b) His songs never last an hour;

    A song, that lasts an hour, is tedious.
    _____
    His songs are never tedious.

    (c) Babies are illogical;

    Nobody is despised who can manage a crocodile;

    Illogical persons are despised.
    _____
    Babies cannot manage crocodiles.

    (Hint **a** : persons who are able to manage a crocodile.

    b : persons who are babies

    c : persons who are despised

    d : persons who are logical)

11 BOOLEAN ALGEBRA

Objectives

After studying this chapter you should

*   be able to use AND, NOT, OR and NAND gates;
*   be able to use combinatorial and switching circuits;
*   understand equivalent circuits;
*   understand the laws of Boolean algebra;
*   be able to simplify Boolean expressions;
*   understand Boolean functions;
*   be able to minimise circuits;
*   understand the significance of half and full adder circuits.

11.0 Introduction

When *George Boole* (1815-186) developed an algebra for logic, little did he realise that he was forming an algebra that has become ideal for the analysis and design of circuits used in computers, calculators and a host of devices controlled by microelectronics. Boole's algebra is physically manifested in electronic circuits and this chapter sets out to describe the building blocks used in such circuits and the algebra used to describe the logic of the circuits.

11.1 Combinatorial circuits

The circuits and switching arrangements used in electronics are very complex but, although this chapter only deals with simple circuits, the functioning of all microchip circuits is based on the ideas in this chapter. The flow of electrical pulses which represent the **binary digits** 0 and 1 (known as **bits**) is controlled by combinations of electronic devices. These **logic gates** act as switches for the electrical pulses. Special symbols are used to represent each type of logic gate.

NOT gate

The NOT gate is capable of reversing the input pulse. The truth table for a NOT gate is as follows:

Input a	Output ~a
0	1
1	0

a —▷o— ~a

*This is a **NOT** gate*

The NOT gate receives an input, either a pulse (1) or no pulse (0) and produces an output as follows :

If input **a** is 1, output is 0;

and if input **a** is 0, output is 1.

AND gate

The AND gate receives two inputs **a** and **b**, and produces an output denoted by $a \wedge b$. The truth table for an AND gate is as follows :

Input a	b	Output $a \wedge b$
0	0	0
0	1	0
1	0	0
1	1	1

a —⊐
b —⊐— a∧b

*This is an **AND** gate*

The only way that the output can be 1 is when **a** AND **b** are both 1. In other words there needs to be an electrical pulse at **a** AND **b** before the AND gate will output an electrical pulse.

OR gate

The OR gate receives two inputs **a** and **b**, and produces an output denoted by $a \vee b$. The truth table for an OR gate is as follows:

Input a	b	Output $a \vee b$
0	0	0
0	1	1
1	0	1
1	1	1

a —⊐
b —⊐— a∨b

*This is an **OR** gate*

The output will be 1 when **a** or **b** or both are 1.

These three gates, NOT, AND and OR, can be joined together to form **combinatorial circuits** to represent Boolean expressions, as explained in the previous chapter.

Example

Use logic gates to represent

(a) $\sim p \vee q$

(b) $(x \vee y) \wedge \sim x$

Draw up the truth table for each circuit

Solution

(a)

p	q	~p	~p∨q
0	0	1	1
0	1	1	1
1	0	0	0
1	1	0	1

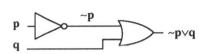

(b)

x	y	x∨y	~x	(x∨y)∧~x
0	0	0	1	0
0	1	1	1	1
1	0	1	0	0
1	1	1	0	0

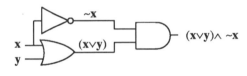

Exercise 11A

Use logic gates to represent these expressions and draw up the corresponding truth tables.

1. $x \wedge (\sim y \vee x)$

2. $a \vee (\sim b \wedge c)$

3. $[a \vee (\sim b \vee c)] \wedge \sim b$

Write down the Boolean expression and the truth table for each of the circuits below.

4.

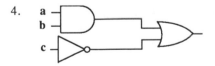

5.

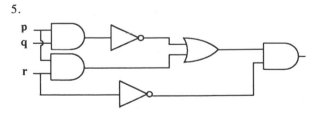

11.2 When are circuits equivalent?

Two circuits are said to be **equivalent** if each produce the same outputs when they receive the same inputs.

Example

Are these two combinatorial circuits equivalent?

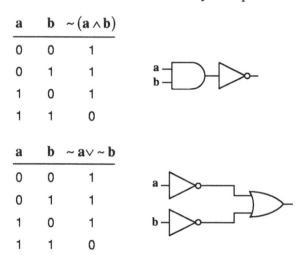

Solution

The truth tables for both circuits will show if they are equivalent :

a	b	$\sim(a \wedge b)$
0	0	1
0	1	1
1	0	1
1	1	0

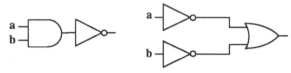

a	b	$\sim a \vee \sim b$
0	0	1
0	1	1
1	0	1
1	1	0

Work through the values in the truth tables for yourself. Since both tables give the same results the two circuits are equivalent. Indeed the two Boolean expressions are equivalent and can be put equal;

i.e. $\sim(a \wedge b) = \ \sim a \vee \sim b$

Exercise 11B

Show if these combinatorial circuits are equivalent by working out the Boolean expression and the truth table for each circuit.

1.

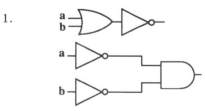

2.

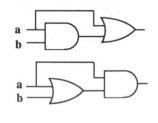

3.
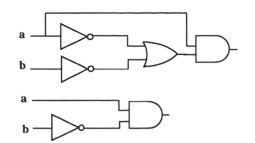

4.
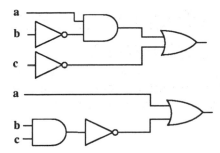

11.3 Switching circuits

A network of switches can be used to represent a Boolean expression and an associated truth table.

Generally the switches are used to control the flow of an electrical current but you might find it easier to consider a switching circuit as a series of water pipes with taps or valves at certain points.

One of the reasons for using switching circuits rather than logic gates is that designers need to move from a combinatorial circuit (used for working out the logic) towards a design which the manufacturer can use for the construction of the electronic circuits.

This diagram shows switches **A**, **B** and **C** which can be **open** or **closed**. If a switch is closed it is shown as a 1 in the following table whilst 0 shows that the switch is open. The **switching table** for this circuit is as follows :

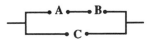

A	B	C	Circuit output
0	0	0	0
0	0	1	1
0	1	0	0
0	1	1	1
1	0	0	0
1	0	1	1
1	1	0	1
1	1	1	1

The table shows that there will be an output (i.e. 1) when **A** AND **B** are 1 OR **C** is 1. This circuit can therefore be represented as

$$\text{(A AND B) OR C}$$

ie. $(A \wedge B) \vee C$

The circuit just considered is built up of two fundamental circuits:

- a **series circuit**, often called an AND gate, $A \wedge B$

————•A•————•B•————

- a **parallel circuit**, often called an OR gate, $A \vee B$.

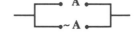

The next step is to devise a way of representing negation. The negation of the truth value 1 is 0 and vice versa, and in switching circuits the negation of a 'closed' path is an 'open' path.

This circuit will always be 'open' whatever the state of **A**. In other words the output will always be 0, irrespective of whether **A** is 1 or 0.

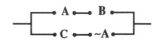

This circuit will always be 'closed' whatever the state of **A**. The output will always be 1 irrespective of whether **A** is 1 or 0.

Example

Represent the circuit shown opposite symbolically and give the switching table.

Solution

The symbolic representation can be built up by considering

the top line of the circuit $(A \wedge B)$

the top bottom of the circuit $(C \wedge \sim A)$.

Combining these gives the result $(A \wedge B) \vee (C \wedge \sim A)$

The table is as follows.

A	B	C	~A	A∧B	C∧~A	(A∧B)∨(C∧~A)
0	0	0	1	0	0	0
0	0	1	1	0	1	1
0	1	0	1	0	0	0
0	1	1	1	0	1	1
1	0	0	0	0	0	0
1	0	1	0	0	0	0
1	1	0	0	1	0	1
1	1	1	0	1	0	1

Activity 1   Make your own circuit

Using a battery, some wire, a bulb and some switches, construct the following circuit.

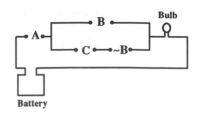

A simple switch can be made using two drawing pins and a paper clip which can swivel to close the switch. The pins can be pushed into a piece of corrugated cardboard or polystyrene.

Using the usual notation of 1 representing a closed switch and 0 representing an open switch, you can set the switches to represent each line of this table:

A	B	C	Output
0	0	0	
0	0	1	
0	1	0	
0	1	1	
1	0	0	
1	0	1	
1	1	0	
1	1	1	

Remember that the '~B' switch is always in the opposite state to the 'B' switch.

Record the output using 1 if the bulb lights up (i.e. circuit is closed) and 0 if the bulb fails to light (i.e. circuit is open)

Represent the circuit symbolically and draw up another table to see if you have the same output.

Exercise 11C

Represent the following circuits by Boolean expresions:

1.

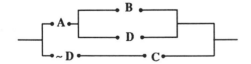

Draw switching circuits for these Boolean expressions:

3. $A \vee (\sim B \wedge C)$

4. $A \wedge ((\sim B \wedge C) \vee (B \wedge \sim C))$

2.

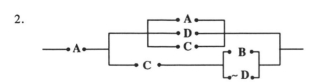

11.4 Boolean algebra

A variety of Boolean expressions have been used but George Boole
was responsible for the development of a complete algebra. In
other words, the expressions follow laws similar to those of the
algebra of numbers.

The operators $\wedge$ and $\vee$ have certain properties similar to those of
the arithmetic operators such as $+$, $-$, $\times$ and $\div$.

(a)   **Associative laws**

$$(a \vee b) \vee c = a \vee (b \vee c)$$

and $(a \wedge b) \wedge c = a \wedge (b \wedge c)$

(b)   **Commutative laws**

$$a \vee b = b \vee a$$

and $a \wedge b = b \wedge a$

(c)   **Distributive laws**

$$a \wedge (b \vee c) = (a \wedge b) \vee (a \wedge c)$$

and $a \vee (b \wedge c) = (a \vee b) \wedge (a \vee c)$

These laws enable Boolean expressions to be simplified and
another law developed by an Englishman, *Augustus de Morgan*
(1806-1871), is useful. He was a contemporary of Boole and
worked in the field of logic and is now known for one important
result bearing his name:-

(d)   **de Morgan's laws**

$$\sim(a \vee b) = \ \sim a \wedge \sim b$$

and $\sim(a \wedge b) = \ \sim a \vee \sim b$

Note: You have to remember to change the connection,
        $\wedge$ changes to $\vee$, $\vee$ changes to $\wedge$.

Two more laws complete the range of laws which are included in
the Boolean algebra.

(e)   **Identity laws**

$$a \vee 0 = a$$

open switch $= 0$

a

and $a \wedge 1 = a$

closed switch $= 1$

(f)   **Complement laws**

$$a \vee \sim a = 1$$

$$\text{and} \quad a \wedge \sim a = 0$$

The commutative law can be developed to give a further result which is useful for the simplification of circuits.

Consider the expressions $a \wedge (a \vee b)$ and the corresponding circuit.

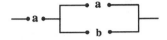

If switch a is open $(a = 0)$ what can you say about the whole circuit? What happens when switch a is closed $(a = 1)$? Does the switch b have any effect on your answers?

The truth table for the circuit above shows that $a \wedge (a \vee b) = a$.

a	b	$a \wedge (a \vee b)$
0	0	0
0	1	0
1	0	1
1	1	1

This result can be extended to more switches. For example

if $a \wedge (a \vee b) = a$

then $a \wedge (a \vee b \vee c) = a$

and $a \wedge (a \vee b \vee c) \wedge (a \vee b) \wedge (b \vee c) = a \wedge (b \vee c).$

The last of these expressions is represented by this circuit:

which can be replaced by the simplified circuit:

Example
Write down a Boolean expression for this circuit. Simplify the expression and draw the corresponding circuit.

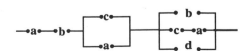

Solution

$$a \wedge b \wedge (a \vee c) \wedge (b \vee (c \wedge a) \vee d)$$
$$a \wedge (a \vee c) \wedge b \wedge (b \vee (c \wedge a) \vee d)$$

Since $\quad a \wedge (a \vee c) = a$

and $\quad b \wedge (b \vee (c \wedge a) \vee d) = b,$

an equivalent expression is $a \wedge b$

and the circuit simplifies to •——• a •——• b •——•

Activity 2   Checking with truth tables

Draw truth tables for the example above to check that

$$a \wedge b \wedge (a \vee c) \wedge (b \vee (c \wedge a) \vee d) = a \wedge b$$

Exercise 11D

Simplify the following and check your answers by drawing up truth tables:

1. $a \vee (\sim a \wedge b)$

2. $a \wedge [b \vee (a \wedge b)] \wedge [a \vee (\sim a \wedge b)]$

3. Simplify the following circuit:

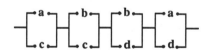

11.5 Boolean functions

In the same way as algebraic functions describe the relationship between the domain, (a set of inputs) and the range (a set of outputs), a **Boolean function** can be described by a **Boolean expression**. For example, if

$$f(x_1, x_2, x_3) = x_1 \wedge (\sim x_2 \vee x_3)$$

then f is the Boolean function and

$$x_1 \wedge (\sim x_2 \vee x_3)$$

is the Boolean expression.

Example

Draw the truth table for the Boolean function defined as

$$f(x_1, x_2, x_3) = x_1 \wedge (\sim x_2 \vee x_3)$$

Solution

The inputs and outputs of this Boolean function are shown in the following table:

x_1	x_2	x_3	$f(x_1,x_2,x_3)$
0	0	0	0
0	0	1	0
0	1	0	0
0	1	1	0
1	0	0	1
1	0	1	1
1	1	0	0
1	1	1	1

It is sometimes necessary to form a function from a given truth table. The method of achieving this is described in the following example.

Example

For the given truth table, form a Boolean function

a	b	c	f(a, b, c)
0	0	0	1
0	0	1	1
0	1	0	0
0	1	1	0
1	0	0	1
1	0	1	0
1	1	0	1
1	1	1	1

Solution

The first stage is to look for the places where f (a,b,c) is 1 and then link them all together with 'OR's. For example, in the last row f (a, b, c) = 1 and this is the row in which a, b and c are all true; i.e. when $a \wedge b \wedge c = 1$.

The output is also 1, (i.e. f $(a, b, c) = 1$) in the 7th row which leads to the combination

$$a \wedge b \wedge \sim c = 1$$

Similarly, for the 5th row

$$a \wedge \sim b \wedge \sim c = 1$$

and for the 2nd row

$$\sim a \wedge \sim b \wedge c = 1$$

and for the 1st row

$$\sim a \wedge \sim b \wedge \sim c = 1$$

All these combinations are joined using the connective $\vee$ to give the Boolean expression

$$(a \wedge b \wedge c) \vee (a \wedge b \wedge \sim c) \vee (a \wedge \sim b \wedge \sim c) \vee (\sim a \wedge \sim b \wedge c) \vee (\sim a \wedge \sim b \wedge \sim c)$$

If the values of a, b and c are as shown in the 1st, 2nd, 5th, 7th and 8th row then the value of f $(a, b, c) = 1$ in each case, and the expression above has a value of 1. Similarly if a, b and c are as shown in the table for which f $(a, b, c) = 0$ then the expression above has the value of 0.

The Boolean function for the truth table is therefore given by

$$f (a, b, c) = (a \wedge b \wedge c) \vee (a \wedge b \wedge \sim c) \vee (a \wedge \sim b \wedge \sim c) \vee (\sim a \wedge \sim b \wedge c) \vee (\sim a \wedge \sim b \wedge \sim c)$$

This is called the **disjunctive normal form** of the function f; the combinations formed by considering the rows with an output value of 1 are joined by the disjunctive connective, OR.

Exercise 11E

Find the disjunctive normal form of the Boolean function for these truth tables:

1.

a	b	f (a, b)
0	0	1
0	1	0
1	0	1
1	1	0

2.

a	b	f (a, b)
0	0	1
0	1	1
1	0	0
1	1	1

3.

x	y	z	f (x, y, z)
0	0	0	1
0	0	1	0
0	1	0	0
0	1	1	0
1	0	0	1
1	0	1	0
1	1	0	0
1	1	1	1

11.6 Minimisation with NAND gates

When designing combinatorial circuits, efficiency is sought by minimising the number of gates (or switches) in a circuit. Many computer circuits make use of another gate called a NAND gate which is used to replace NOT AND, thereby reducing the number of gates.

The NAND gate receives inputs **a** and **b** and the output is denoted by $a \uparrow b$.

The symbol used is

The truth table for this is

a	b	$a \uparrow b$
0	0	1
0	1	1
1	0	1
1	1	0

The NAND gate is equivalent to

Note that, by de Morgan's law, $\sim (a \wedge b) = \sim a \vee \sim b$.

Example

Use NAND gates alone to represent the function

$$f(a, b, c, d) = (a \wedge b) \vee (c \wedge d)$$

Solution

The use of NAND gates implies that there must be negation so the function is rewritten using de Morgan's Laws:

$$(a \wedge b) \vee (c \wedge d) = \sim [(\sim a \vee \sim b) \wedge (\sim c \vee \sim d)]$$

The circuit consisting of NAND gates is therefore as follows:

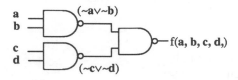

Example

Design combinatorial circuits to represent (a) the negation function $f(x) = \sim x$ and (b) the OR function $f(x, y) = x \vee y$.

Solution

(a) $\sim x = \sim (x \vee x)$

$\qquad = \sim x \wedge \sim x$

$\qquad = x \uparrow x$

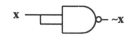

(b) $x \vee y = \sim (\sim x \wedge \sim y)$

$\qquad = \sim x \uparrow \sim y$

$\qquad = (x \uparrow x) \uparrow (y \uparrow y)$

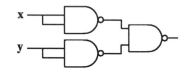

Exercise 11F

Design circuits for each of the following using only NAND gates.

1. $a \wedge b$

2. $a \wedge \sim b$

3. $(\sim a \wedge \sim b) \vee \sim b$

11.7 Full and half adders

Computers turn all forms of data into **binary digits** , (0 s and 1 s), called bits, which are manipulated mathematically. For example the number 7 is represented by the binary code 00000111 (8 bits are used because many computers use binary digits in groups of 8, for example, ASCII code). This section describes how binary digits can be added using a series of logic gates. The basic mathematical operation is **addition** since

> **subtraction**   is the addition of negative numbers,
>
> **multiplication**   is repeated addition,
>
> **division**   is repeated subtraction.

When you are adding two numbers there are two results to note for each column; the entry in the answer and the carrying figure.

$$
\begin{array}{r}
254 \\
178 \\
\hline
432 \\
\hline
11
\end{array}
$$

When 4 is added to 8 the result is 12, 2 is noted in the answer and the digit 1 is carried on to the next column.

When adding the second column the carry digit from the first column is included, i.e. $5+7+1$, giving yet another digit to carry on to the next column.

Half adder

The half adder is capable of dealing with two inputs, i.e. it can only add two bits, each bit being either 1 or 0.

a	b	Carry bit	Answer bit
0	0	0	0
0	1	0	1
1	0	0	1
1	1	1	0

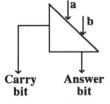

Carry bit   Answer bit

Activity 3   Designing the half adder circuit

The next stage is to design a circuit which will give the results shown in the table above.

The first part of the circuit is shown opposite; complete the rest of the circuit which can be done with a NOT gate, an OR gate and an AND gate to give the answer bit.

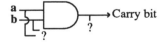

Full Adder

A half adder can only add two bits; a full adder circuit is capable of including the carry bit in the addition and therefore has three inputs.

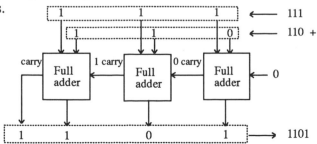

Activity 4   Full adder truth table

Complete this truth table for the full adder.

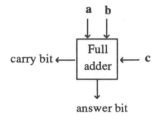

Inputs				
a	b	c	Carry bit	Answer bit
0	0	0	0	0
0	0	1	0	1
0	1	0		
	etc ↓			
1	1	1		

The circuit for a full adder is, in effect, a combination of two half adders.

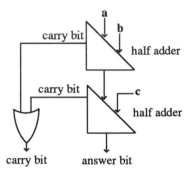

If you think about it, the carry bit of the full adder must be 1 if either of the two half adders shown gives a carry bit of 1 (and in fact it is impossible for both those half adders to give a carry bit of 1 at the same time). Therefore the two carry bits from the half adders are fed into an OR gate to give an output equal to the carry bit of the full adder.

The circuit for a full adder consists, therefore, of two half adders with the carry bits feeding into an OR gate as follows:

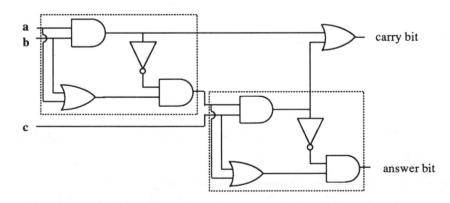

The dotted lines enclose the two half adders with the whole circuit representing a full adder.

Activity 5   NAND half adder

Draw up a circuit to represent a half adder using only NAND gates.

11.8 Miscellaneous Exercises

1. Use logic gates to represent these expressions and draw up the corresponding truth tables:

   (a) $\sim[(a \wedge b) \vee c]$

   (b) $(a \wedge b) \vee \sim c$

   (c) $\sim c \wedge [(a \wedge b) \vee \sim (a \wedge c)]$

2. Write down the Boolean expression and the truth table for each of these circuits:

   (a)
   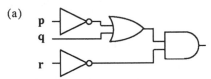

   (b)
   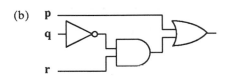

   (c)
   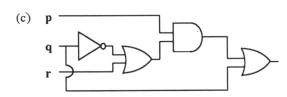

3. Write down the Boolean expressions for these circuits :

   (a)
   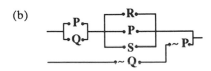

   (b)
   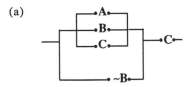

   (c)
   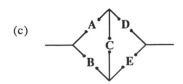

4. Draw switching circuits for these Boolean expressions:

   (a) $(B \wedge C) \vee (C \wedge A) \vee (A \wedge B)$

   (b) $\left[A \wedge ((B \wedge \sim C) \vee (\sim B \wedge C))\right] \vee (\sim A \wedge B \wedge C)$

5. Draw the simplest switching circuit represented by this table:

P	Q	R	S	Output
1	1	1	1	1
0	1	1	1	1
1	1	0	1	1
0	1	0	1	1

6. A burglar alarm for a house is controlled by a switch. When the switch is on, the alarm sounds if either the front or back doors or both doors are opened. The alarm will not work if the switch is off. Design a circuit of logic gates for the alarm and draw up the corresponding truth table.

*7. A gallery displaying a famous diamond uses a special Security Unit to protect access to the Display Room (D). The diagram below shows the layout of the system.

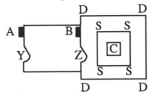

The display cabinet (C) is surrounded by a screen of electronic eyes (S).

Access to the display room is through doors (Y), (Z). Boxes (A), (B) are used in the system. The following persons are involved in the system :

>   Manager,
>   Deputy Manager,
>   Chief Security Officer.

The Display Room is opened as follows :

The Unit must be activated at box A.

Door (Y) is opened by any two of the above persons at box A.

Box B is activated by the Manager and Deputy Manager together.

The screen (S) is activated by the Chief Security Officer alone at box B only.

Door (Z) can only be opened once the screen (S) is activated.

Draw a circuit of logic gates required inside the Unit to operate it. Ensure your diagram is documented.

(AEB)

8. Simplify the following expressions and check your answer by drawing up truth tables.

    (a) $(a \wedge b \wedge c) \vee (\sim a \wedge b \wedge c)$

    (b) $a \vee (\sim a \wedge b \wedge c) \vee (\sim a \wedge b \wedge \sim c)$

    (c) $(p \wedge q) \vee (\sim p \vee \sim q) \wedge (r \vee s)$

9. Find the disjunctive normal form of this function; simplify and draw the combinatorial circuit.

a	b	c	f (a,b,c)
0	0	0	0
0	0	1	0
0	1	0	0
0	1	1	1
1	0	0	0
1	0	1	1
1	1	0	1
1	1	1	1

10. Design a circuit representing $\sim a \vee b$ using NAND gates.

*11. Write a computer program or use a spreadsheet that outputs a truth table for a given Boolean expression.

12. (a) Establish a truth table for the Boolean function $f(x_1, x_2, x_3) = (\sim x_1 \vee x_2) \wedge (\sim x_3 \vee x_2)$.

    (b) Design a circuit using as few AND, OR and NOT gates as possible to model the function in (a).

13. (a) Show, by constructing truth tables or otherwise, that the following statements are equivalent.

    $p \Rightarrow q$ and $\sim (\sim (p \wedge q) \wedge p)$.

    (b) With the aid of (a), or otherwise, construct combinatorial circuits consisting only of NAND gates to represent the functions

    $f(x, y) = x \Rightarrow y$ and $g(x, y) = \sim (x \Leftrightarrow y)$.

12 CRITICAL PATH ANALYSIS

Objectives

After studying this chapter you should

- be able to construct activity networks;
- be able to find earliest and latest starting times;
- be able to identify the critical path;
- be able to translate appropriate real problems into a suitable form for the use of critical path analysis.

12.0 Introduction

A complex project must be well planned, especially if a number of people are involved. It is the task of management to undertake the planning and to ensure that the various tasks required in the project are completed in time.

Operational researchers developed a method of scheduling complex projects shortly after the Second World War. It is sometimes called **network analysis**, but is more usually known as **critical path analysis** (CPA). Its virtue is that it can be used in a wide variety of projects, and was, for example, employed in such diverse projects as the Apollo moonshot, the development of Concorde, the Polaris missile project and the privatisation of the electricity and water boards. Essentially, CPA can be used for any multi-task complex project to ensure that the complete scheme is completed in the minimum time.

Although its real potential is for helping to schedule complex projects, we will illustrate the use of CPA by applying it to rather simpler problems. You will often be able to solve these problems without using CPA, but it is an understanding of the concepts involved in CPA which is being developed here.

12.1 Activity networks

In order to be able to use CPA, you first need to be able to form what is called an **activity network**. This is essentially a way of illustrating the given project data concerning the tasks to be completed, how long each task takes and the constraints on the order in which the tasks are to be completed. As an example, consider the activities shown below for the construction of a garage.

	activity	duration (in days)
A	prepare foundations	7
B	make and position door frame	2
C	lay drains, floor base and screed	15
D	install services and fittings	8
E	erect walls	10
F	plaster ceiling	2
G	erect roof	5
H	install door and windows	8
I	fit gutters and pipes	2
J	paint outside	3

Clearly, some of these activities cannot be started until other activities have been completed. For example

>   activity G - erect roof

cannot begin until

>   activity E - erect walls

has been completed. The following table shows which activities must precede which.

>   D   must follow E
>   E   must follow A and B
>   F   must follow D and G
>   G   must follow E
>   H   must follow G
>   I   must follow C and F
>   J   must follow I.

We call these the **precedence relations**.

All this information can be represented by the network shown below.

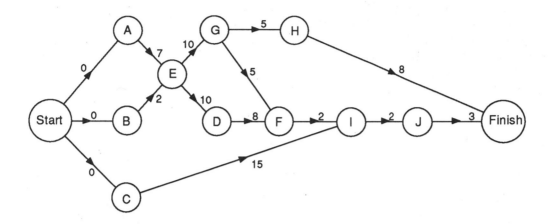

In this network

each **activity** is represented by a **vertex**;

joining vertex X to vertex Y shows that
activity X must be **completed** before Y can be started;

the **number** marked on each arc shows the **duration** of the
activity from which the arc starts.

Note the use of 'arc' here to mean a directed edge.
Sometimes we can easily form the activity network, but not
always, so we need to have a formal method. First try the
following activity.

Activity 1   Making a settee

A furniture maker is going to produce a new wooden framed settee
with cloth-covered foam cushions. These are the tasks that have to
be done by the furniture maker and his assistants and the times
they will take :

	activity	time in days
A	make wooden arms and legs	3
B	make wooden back	1
C	make wooden base	2
D	cut foam for back and base	1
E	make covers	3
F	fit covers	1
G	put everything together	1

Each activity can only be undertaken by one individual.

The following list gives the order in which the jobs must be done:

B   must be after   C

A   must be after   B and C

D   must be after   B and C

E   must be after   D

F   must be after   E

G   must be after   A, B, C, D, E and F

Construct an appropriate activity network to illustrate this information.

12.2 Algorithm for constructing activity networks

For simple problems it is often relatively easy to construct activity networks but, as the complete project becomes more complex, the need for a formal method of constructing activity networks increases. Such an algorithm is summarised below.

		Original vertices	Shadow vertices
Start	Write down the original vertices and then a second copy of them alongside, as illustrated on the right. If activity Y must follow activity X draw an arc from original vertex Y to shadow vertex X. (In this way you construct a **bipartite graph**.)	A ● ○ A B ● ○ B C ● ○ C ⋮ ⋮ X ● ○ X Y ● ○ Y	
Step 1	Make a list of all the original vertices which have **no** arcs incident to them.		
Step 2	Delete all the vertices found in Step 1 and their corresponding shadow vertices and all arcs incident to these vertices.		
Step 3	Repeat Steps 1 and 2 until all the vertices have been used.		

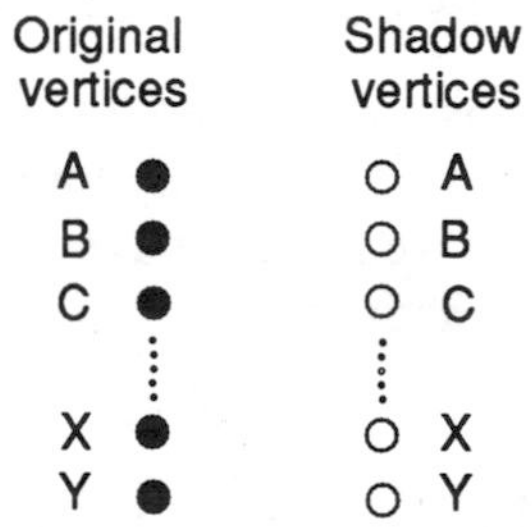

The use of this algorithm will be illustrated using the first case study, constructing a garage, from Section 12.1.

The precedence relations are:

D must follow E

E must follow A and B

F must follow D and G

G must follow E

H must follow G

I must follow C and F

J must follow I

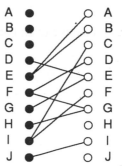

These are illustrated opposite.

Applying the algorithm until all vertices have been chosen is shown below.

Step 1 - original vertices with no arcs

A, B, C

Step 2 - delete all arcs incident on A, B, C and redraw as shown

Step 3 - repeat iteration

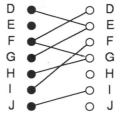

Step 1 - original vertices with no arcs

E

Step 2 - delete all arcs incident on E and redraw as shown

Step 3 - repeat iteration

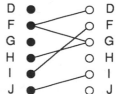

Step 1 - original vertices with no arcs

D, G

Step 2 - delete all arcs incident on D, G and redraw as shown

Step 3 - repeat iteration

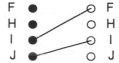

Step 1 - original vertices with no arcs

F, H

Step 2 - delete all arcs incident on F, H and redraw as shown

Step 3 - repeat iteration

Step 1 - original vertices with no arcs

I

Step 2 - delete all arcs incident on I and redraw as shown

Step 3 - stop as all vertices have been chosen

So the vertices have been chosen in the following order:

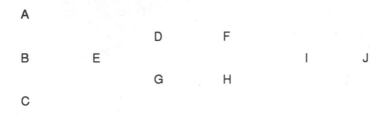

The activity diagram as shown belowcan now be drawn.

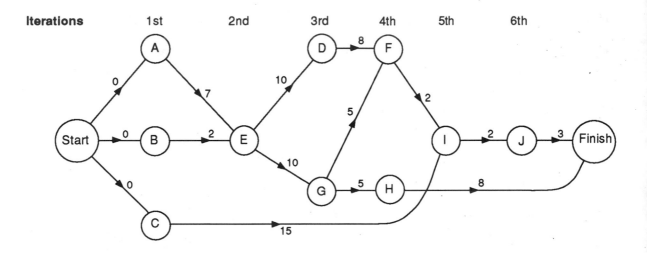

From the 'start' vertex, draw arcs to A, B and C, the first iteration vertices, putting zero on each arc. In the original bipartite graph the shadow vertex A was joined to the original vertex E - so join A to E. Similarly join B to E and C to I.

Indicate the duration of the activity on any arc coming **from** the vertex representing the activity.

Continue in this way and complete the activity network with a 'finish' vertex into which any free vertices lead, again indicating the duration of the activity on the arc.

Note that the duration of the activity is shown on every arc **coming** from the vertex representing the activity. (So, for example, arc ED and arc EG are both given 10.)

Exercise 12A

1. Use the algorithm to find the activity network for the problem in Activity 1.

2. Suppose you want to redecorate a room and put in new self-assembly units. These are the jobs that need to be done, together with the time each takes:

activity	time (in hrs)	preceded by
paint woodwork (A)	8	-
assemble units (B)	4	-
fit carpet (C)	5	hang wallpaper paint woodwork
hang wallpaper (D)	12	paint woodwork
hang curtains (E)	2	hang wallpaper paint woodwork

Complete an activity network for this problem.

3. The Spodleigh Bicycle Company is getting its assembly section ready for putting together as many bicycles as possible for the Christmas market. This diagram shows the basic components of a bicycle.

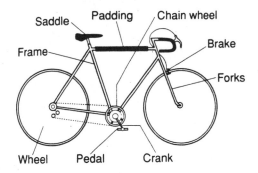

Putting together a bicycle is split up into small jobs which can be done by different people. These are:

	activity	time (mins)
A	preparation of the frame	9
B	mounting and aligning the front wheel	7
C	mounting and aligning the back wheel	7
D	attaching the chain wheel to the crank	2
E	attaching the chain wheel and crank to the frame	2
F	mounting the right pedal	8
G	mounting the left pedal	8
H	final attachments such as saddle, chain, stickers, etc.	21

The following chart shows the order of doing the jobs.

B must be after A

C must be after A

D must be after A

E must be after D

F must be after D and E

G must be after D and E

H must be after A, B, C, D, E, F and G

Draw an activity network to show this information.

4. An extension is to be built to a sports hall. Details of the activities are given below.

	activity	time (in days)
A	lay foundations	7
B	build walls	10
C	lay drains and floor	15
D	install fittings	8
E	make and fit door frames	2
F	erect roof	5
G	plaster ceiling	2
H	fit and paint doors and windows	8
I	fit gutters and pipes	2
J	paint outside	3

Some of these activities cannot be started until others have been completed:

B must be after C

C must be after A

D must be after B

E must be after C

F must be after D and E

G must be after F

H must be after G

I must be after F

J must be after H

Complete an activity network for this problem.

12.3 Critical path

You have seen how to construct an activity network. In this
section you will see how this can be used to find the **critical path**.
This will first involve finding the **earliest** possible start for each
activity, by going **forwards** through the network. Secondly, the
latest possible start time for each activity is found by going
backwards through the network. Activities which have **equal**
earliest and latest start time are on the **critical path**. The
technique will be illustrated using the 'garage construction'
problem from Sections 12.1 and 12.2.

The activity network for this problem is shown below, where
sufficient space is made at each activity node to insert two
numbers.

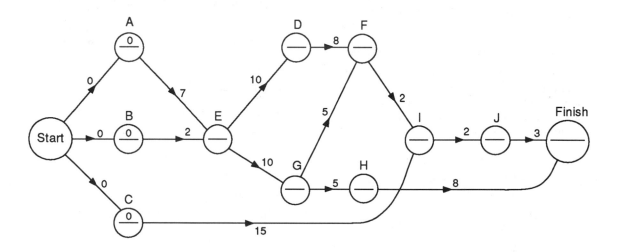

The numbers in the top half of each circle will indicate the earliest
possible starting time. So, for activities A, B and C, the number
zero is inserted.

Moving forward through the network, the activity E is reached
next. Since both A and B have to be completed before E can be
started, the earliest start time for E is 7. This is put into the top
half of the circle at E. The earliest times at D and G are then both
17, and for H, 22. Since F cannot be started until both D and G are
completed, its earliest start time is 25, and consequently, 27 for I.
The earliest start time for J is then 29, which gives an earliest
completion time of 32.

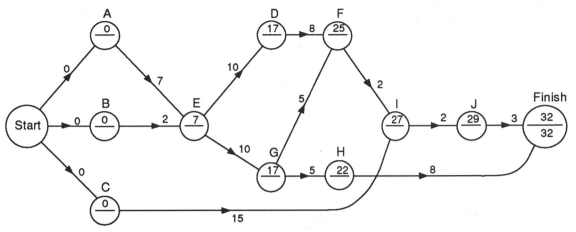

Since 32 is the earliest possible completion time, it is also assumed to be the completion time in order to find the latest possible start times. So 32 is also put in the lower half of the 'finish' circle. Now working backwards through the network, the latest start times for each activity are as follows:

J $32 - 3 = 29$

I $29 - 2 = 27$

F $27 - 2 = 25$

H $32 - 8 = 24$

D $25 - 8 = 17$

G   the minimum of $25 - 5 = 20$ and $24 - 5 = 19$

E   the minimum of $17 - 10 = 7$ and $19 - 10 = 9$

A $7 - 7 = 0$

B $7 - 2 = 5$

C $27 - 15 = 12$

This gives a completed network as shown below.

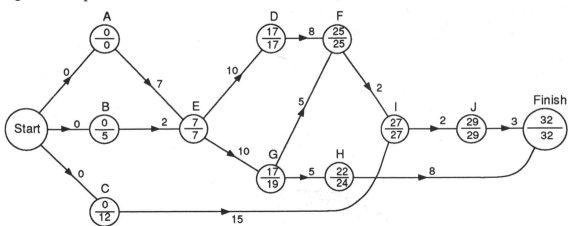

The vertices with equal earliest and latest starting times define the **critical path**. This is clearly seen to be

A E D F I J.

Another way of identifying the critical path is to define the

float time = latest start time − earliest start time.

The information for the activities can now be summarised in the table below.

activity	start times earliest	latest	float	
A	0	0	0	←
B	0	5	5	
C	0	12	12	
E	7	7	0	←
D	17	17	0	←
G	17	19	2	
F	25	25	0	←
H	22	24	2	
I	27	27	0	←
J	29	29	0	←

So now you know that if there are enough workers the job can be completed in 32 days. The activities on the critical path (i.e. those with zero float time) must be started punctually; for example, A must start immediately, E after 7 days, F after 25 days, etc. For activities with a non-zero float time there is scope for varying their start times; for example activity G can be started any time after 17, 18 or 19 days' work. Assuming that all the work is completed on time, you will see that this does indeed give a working schedule for the construction of the garage in the minimum time of 32 days. However it does mean, for example, that on the 18th day activities D and C will definitely be in progress and G may be as well. The solution could well be affected if there was a limit to the number of workers available, but you will consider that sort of problem in the next chapter.

Is a critical path always uniquely defined?

Activity 2   Bicycle construction

From the activity network for Question 3 in Exercise 12A find the critical path and the possible start times for all the activities in order to complete the job in the shortest possible time.

Exercise 12B

1. Find the critical paths for each of the activity networks shown below.

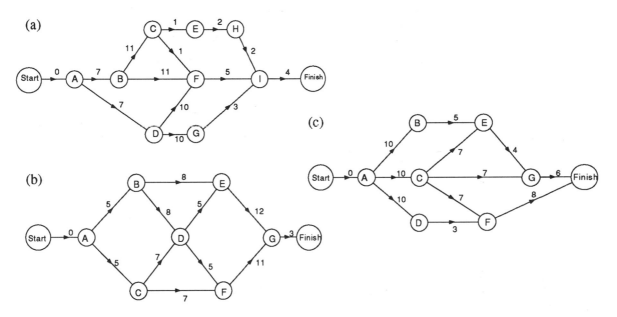

(a)

(b)

(c)

2. Find the critical path for the activity network in Question 4, Exercise 12A.

3. Your local school decides to put on a musical. These are the many jobs to be done by the organising committee, and the times they take:

A	make the costumes	6 weeks
B	rehearsals	12 weeks
C	get posters and tickets printed	3 weeks
D	get programmes printed	3 weeks
E	make scenery and props	7 weeks
F	get rights to perform the musical	2 weeks
G	choose cast	1 week
H	hire hall	1 week
I	arrange refreshments	1 week
J	organise make-up	1 week
K	decide on musical	1 week
L	organise lighting	1 week
M	dress rehearsals	2 days
N	invite local radio/press	1 day
P	choose stage hands	1 day
Q	choose programme sellers	1 day
R	choose performance dates	$\frac{1}{2}$ day
S	arrange seating	$\frac{1}{2}$ day
T	sell tickets	last 4 weeks
V	display posters	last 3 weeks

(a) Decide on the precedence relationships.

(b) Construct the activity network.

(c) Find the critical path and minimum completion time.

12.4 Miscellaneous Exercises

1. Consider the following activity network, in which the vertices represent activities and the numbers next to the arcs represent time in days.

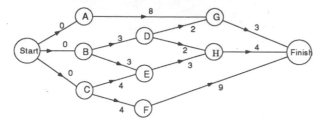

(a) Assuming that an unlimited number of workers is available, write down:

  (i) the minimum completion time of the project if an unlimited number of workers is available;

  (ii) the corresponding critical path.

(b) Find the float time of activity E.

2. A project consists of ten activities, A-J. The duration (in days) of each activity, and the activities preceding each of them, are as follows:

No. of workers

activity		duration	preceding activities
A	2	10	-
B	2	4	-
C	2	8	B
D	2	6	C
E	2	8	I
F	2	5	-
G	2	10	A, D
H	2	2	G
I	2	4	-
J	2	10	D, F, I

Using the algorithms in Section 12.2,

(a) construct an activity network for this project;

(b) find a critical path in this activity network;

(c) find the latest starting time for each activity.

3. A project consists of eight activities whose durations are as follows:

activity	A	B	C	D	E	F	G	H
duration	4	4	3	5	4	1	6	5

The precedence relations are as follows:

  B must follow A

  D must follow A and C

  F must follow C and E

  G must follow C and E

  H must follow B and D.

(a) Draw an activity network in which the activities are represented by vertices.

(b) Find a critical path by inspection, and write down the earliest and latest starting times for each activity.

4. The eleven activities A to K which make up a project are subject to the following precedence relations.

No. of nodes.

preceding activities	activity		duration
C, F, J	A	2	7
E	B	3	6
-	C	3	9
B, H	D	2	7
C, J	E	3	3
-	F	4	8
A, I	G	3	4
J	H	2	9
E, F	I	2	9
-	J	3	7
B, H, I	K	4	5

(a) Construct an activity network for the project.

(b) Find:

  (i) the earliest starting time of each activity in the network;

  (ii) the latest starting time of each activity.

(c) Calculate the float of each activity, and hence determine the critical path.

5. The activities needed to replace a broken window pane are given below.

activity		duration (in mins)	preceding activities
A	order glass	10	-
B	collect glass	30	A
C	remove broken pane	15	B, D
D	buy putty	20	-
E	put putty in frame	3	C
F	put in new pane	2	E
G	putty outside and smooth	10	F
H	sweep up broken glass	5	C
I	clean up	5	all

(a) Construct an activity network.

(b) What is the minimum time to complete the replacement?

(c) What is the critical path?

6. Write the major activities, duration time and precedence relationship for a real life project with which you are involved. Use the methods in this chapter to find the critical path for your project.

7. Consider the following activity network, in which the vertices represent activities and the the numbers next to the arcs represent time in weeks:

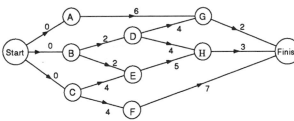

(a) Write down the minimum completion time of the project, if an unlimited number of workers is available, and the corresponding critical path.

(b) Find the float times of activities D and B.

8. A firm of landscape gardeners is asked to quote for constructing a garden on a new site. The activities involved are shown in the table.

activity		duration (in days)	preceding activities
A	prepare site	2	-
B	build retaining wall for patio	3	A
C	lay patio * (see below)	4	A
D	lay lawn	1	A
E	lay paths	3	A B
F	erect pergola, trellis, etc.	1	A B D G
G	prepare flower beds and border	1	A B D
H	plant out	3	A B D G
I	clean up	1	all

* Note also that the patio cannot begin to be laid until 2 days after the start of the building of the retaining wall.

(a) Construct an activity network for this problem.

(b) Find the earliest and latest start time for each activity, state the minimum time for completion of the work and identify the critical path.

(c) Which activities have the greatest float time?

(AEB)

9. At 4.30 pm one day the BBC news team hear of a Government Minister resigning. They wish to prepare an item on the event for that evening's 6 o'clock news. The table below lists the jobs needed to prepare this news item, the time each job takes and the constraints on when the work can commence.

Job		Time needed (in minutes)	Constraints
A	Interview the resigning Minister	15	Starts at 4.30 pm
B	Film Downing St.	20	None
C	Get reaction from regions	25	Cannot start until A and B are completed
D	Review possible replacements	40	Cannot start until B is completed
E	Review the Minister's career	25	Cannot start until A is completed
F	Prepare film for archives	20	Cannot start until C and E are completed
G	Edit	20	Cannot start until A, B, C, D, E and F are completed

(a) Construct an activity network for this problem and, by finding the critical path in your network, show that the news item can be ready before 6.00 pm that day.

(b) If each of the jobs A, B, C, D, E and F needs a reporter, and once a reporter has started a job that same reporter alone must complete i;, explain how three reporters can have the news item ready before 6.00 pm, but that two reporters cannot.   (AEB)

13 SCHEDULING

Objectives

After studying this chapter you should

*   be able to apply a scheduling algorithm to Critical Path
    Analysis problems;

*   appreciate that this does not always produce the optimum
    solution;

*   be able to design methods for solving packing problems;

*   be able to use the branch-and-bound method for solving the
    knapsack problem.

13.0 Introduction

In the previous chapters it was possible to find the critical path for
complex planning problems, but no consideration was given to
how many workers would be available to undertake the activities,
or indeed to how many workers would be needed for each activity.

You will see how scheduling methods can be applied to Critical
Path Analysis problems but, importantly, it will be shown that such
methods do not necessarily give the **optimal** solution every time.

You will also see how this scheduling problem is related to bin-
filling problems, and to a similar problem called the knapsack
problem in which the items carried not only have particular
weight, but also have an appropriate **value**.

Activity 1

Find the critical path for the activity network shown below.

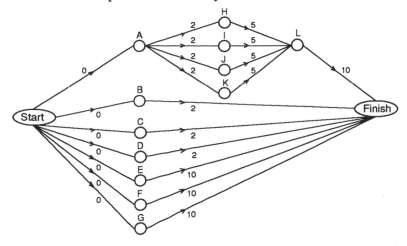

Suppose that each activity can be undertaken by a single worker, and that there are just 2 workers available. Also, that once an activity has been started by a worker, it must be completed by that same worker with no stoppages.

Design a schedule for these two workers so that the complete project is completed in the **minimum** time possible.

Does the minimum completion time depend on the number of workers available?

13.1 Scheduling

As in Activity 1, the following **operating rules** will be assumed:

1.	Each activity requires only **one** worker.
2.	No worker may be idle if there is an activity that can be started.
3.	Once a worker starts an activity, it must be continued by that worker until it is completed.

The **objective** will be to:

> *'Complete the project as soon as possible with the available number of workers.'*

The main example from Chapter 12, which related to the construction of a garage, will be used to illustrate the problem.

The activity network, earliest and latest starting times, and the critical path (bold line), are shown below.

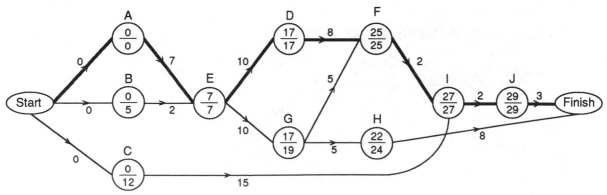

Suppose there are two workers available for the complete project. What is needed is a set procedure in order to decide who does what.

Can you think what would be a suitable procedure for allocating workers to activities?

The method that will be adopted can be summarised as follows.

> At any stage, when a worker becomes free, consider all the activities which have not yet been started but which can now be started. Assign to the worker the most 'critical' one of these (i.e. the one whose latest starting time is the smallest). If there are no activities which can be started at this stage you may have to wait until the worker can be assigned a job.

Using this as a basis for decisions, the solution shown opposite is obtained.

Note that worker 1 completes all the activities on the critical path, though, at time $t = 17$, workers 1 and 2 could have swopped over.

Since the whole project is completed in time 32 days, which you already know to be the minimum completion time, you can be assured that this method has produced an optimum solution. However, this is not always the case, as you will see in the next example.

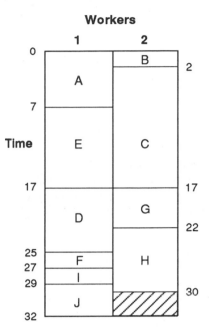

Example

The problem in Activity 1 has the following activity network and critical path. Use the method above to schedule four workers for this project.

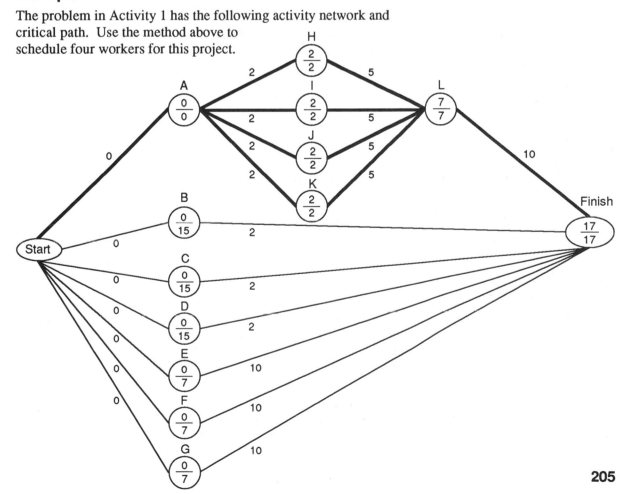

Solution

Applying the method as before results in the schedule shown in the first diagram opposite. This gives a completion time of $t = 25$.

Is this solution optimal?

It should not take too long to find a schedule which completes the project in time 17. A possible solution is shown in the second diagram .

As no worker is ever idle, and they all finish at time 17, this must be an optimum solution - you cannot do better! So the algorithm does not always produce the optimum solution. Currently no procedure exists which always guarantees to give the optimum scheduling solution, except the method of exhaustion where every possible schedule is tested.

Workers

1	2	3	4
A			
H	E	F	G
I			B
C	J	K	D
L			

Workers

1	2	3	4
A	B	C	D
H	I	J	K
E	F	G	L

Activity 2

A possible revision of the method is to :

> **Evaluate** for each activity, the sum of the earliest and latest starting times, and **rank** the activities in ascending order according to this sum.

Then activities are assigned according to this ranking, taking the precedence relations into account. Use this method to schedule the project above, again using four workers. Does it produce the optimal solution?

Activity 3

Design your own method of scheduling. Try it out on the two examples above.

Exercise 13A

1. Use the first scheduling method to find a solution to Question 3 in Section 12.4, using two workers. Does this produce an optimum schedule?

2. Find a schedule for the problem given in Question 3, Exercise 12A, using 2 workers and the two scheduling algorithms given in this section.

   Does either of these methods provide an optimal solution?

3. A possible modification to the method in this section is as follows :

   'Evaluate for each activity the **product** of the earliest and latest starting times, and rank the activities in ascending order according to these numbers. Assign activities using this ranking, taking the precedence relations into account.'

   Use this method to find possible schedules for the garage construction problem in this section. Does this method always give an optimum solution?

13.2 Bin packing

In the previous section, you saw how to schedule activities for a given number of workers in order to complete the project in minimum time. In this section, the problem is turned round and essentially asks for the minimum number of workers required to complete the project within a given time. It will be assumed that there are no precedence relations. The difficulties will be illustrated in the following problem.

A project consists of the following activities (with no precedence relations):

Activity	A	B	C	D	E	F	G	H	I	J	K
Duration (in days)	8	7	4	9	6	9	5	5	6	7	8

What is the minimum number of workers required to complete the project in 15 days?

Find a lower bound to the minimum number of workers needed.

Activity 4

Show that there is a solution to this problem which uses only five workers.

You should obtain a solution without too much difficulty. It is more difficult, however, to find an **algorithm** to solve such problems. The problem considered here is one involving **bin packing**. If you replace workers by bins, each having a maximum capacity of 15 units, the problem is to use the minimum number of bins.

Think about how a precise method could be designed to solve problems of this type.

One possible method is known as **first-fit packing** :

> Number the bins, then always place the next item in the lowest numbered bin which can take that item.

Applying this method to the problem above gives the following solution.

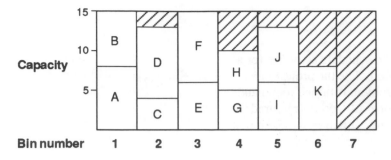

Using this method, six bins are needed, but you should have found a solution in Activity 4 which needs just five bins. So this method does not necessarily produce the optimum solution.

How can the first-fit method be improved?

Looking at the way the method works, it seems likely that it might be improved by just reordering the items into decreasing order of size, so that the items of largest size are packed first. Then you have the **first-fit decreasing method** :

> 1. Reorder the items in decreasing order of size.
>
> 2. Apply the first fit procedure to the reordered list.

You will see how this works using the same problem as above. First reorder the activities in decreasing size.

Activity	D	F	A	K	B	J	E	I	G	H	C
Duration (in days)	9	9	8	8	7	7	6	6	5	5	4

and then apply the method to give the solution below.

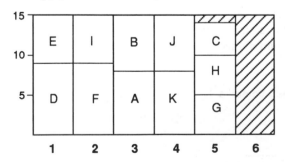

This clearly gives an optimal solution.

Will this method always give an optimal solution?

Bin-packing problems occur in a variety of contexts. As you have already seen, one context is that of determining the minimum number of workers to complete a project in a specified time period. Other examples occur in :

Plumbing   in which it is required to minimise the number of pipes of standard length required to cut a specified number of different lengths of pipe.

Advertising on television, in which case the bins are the standard length breaks between programmes, with the problem of trying to pack a specified list of adverts into the smallest number of breaks.

Example

A builder has piping of standard length 12 metres.

The following sections of various lengths are required

Section	A	B	C	D	E	F	G	H	I	J	K	L
Length (in metres)	2	2	3	3	3	3	4	4	4	6	7	7

Find a way of cutting these sections from the standard 12 m lengths so that a minimum number of lengths is used. Use

    (a)   first-fit method,

    (b)   first-fit decreasing method,

    (c)   trial and error,

to find a solution.

Solution

(a) First-fit method

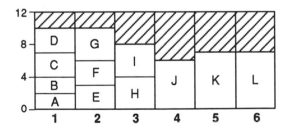

(b) First-fit decreasing method

Section	L	K	J	I	H	G	F	E	D	C	B	A
Length (in metres)	7	7	6	4	4	4	3	3	3	3	2	2

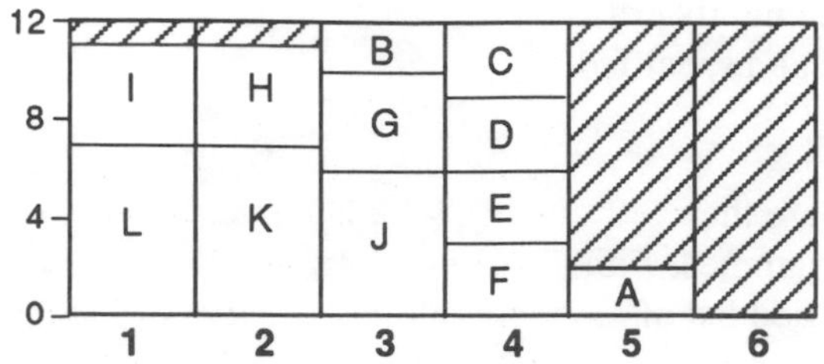

(c) Trial and Error

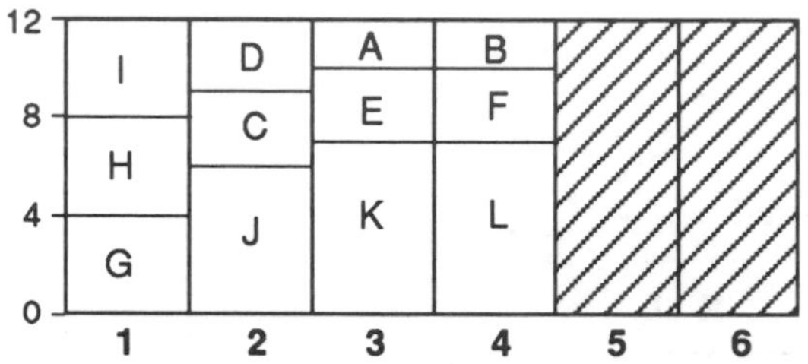

Note that even the first-fit decreasing method does not necessarily give the optimum solution, as shown above. An indication of the number of bins required can be obtained by evaluating

$$\frac{\text{sum of all sizes}}{\text{bin size}}$$

and noting the smallest integer that is greater than (or equal to) this number. This integer is a **lower bound** to the number of bins required. However you cannot always obtain a solution with this number.

Example

Find an optimum solution for fitting items of size

$$7, \ 6, \ 6, \ 6, \ 4, \ 3$$

into bins of size 11.

Solution

Noting that

$$\frac{\text{sum of all sizes}}{\text{bin size}} = \frac{7+6+6+6+4+3}{11} = \frac{32}{11} = 2\tfrac{10}{11},$$

it can be see that three is a lower bound to the number of bins required. But it is clear that four bins will in fact be needed and that no solution exists using just three bins.

A possible solution is given opposite.

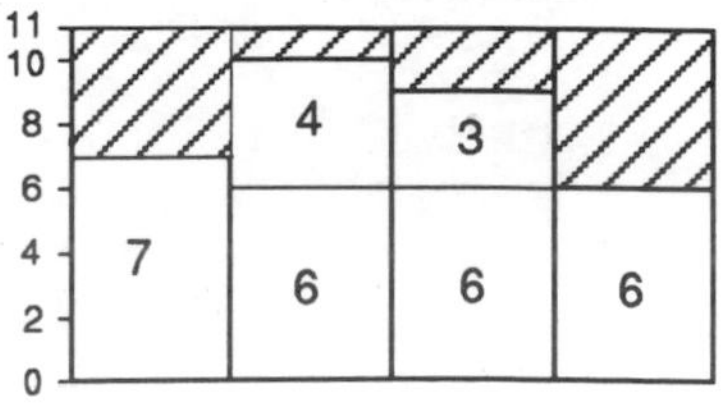

Exercise 13B

1. A project consists of eight activities whose durations are as follows. There are no precedence relations.

Activity	A	B	C	D	E	F	G	H
Duration (hours)	1	2	3	4	4	3	2	1

It is required to find the minimum number of workers needed to finish the project in 5 hours.

Find the answers to this problem given by

(a) the first-fit packing method;

(b) the first-fit decreasing method.

2. A plumber uses pipes of standard length 10 m and wishes to cut out the following lengths

Length (m)	10	9	8	7	6	5	4	3	2	1
Number	0	0	2	3	1	1	0	2	3	0

Use the first-fit decreasing method to find how many standard lengths are needed to meet this order. Does this method give an optimum solution? If not, find an optimum solution.

3. Determine the minimum number of sheets of metal required, 10 m by 10 m, to meet the following order, and how they should be cut. (Assume no wastage in cutting.)

Size	Number
$3 \times 1\,m^2$	60
$4 \times 2\,m^2$	49
$7 \times 5\,m^2$	12

Develop a **general** method of solving 2-dimensional packing problems of this type.

*13.3 Knapsack problem

For the bin-filling problem, the aim was to pack items of different sizes into a minimum number of bins. Now suppose that there is just **one** bin, but that each item has a value associated with it. Thus the question is what items should be packed in order to maximise the total value of the items packed. The problem is known as the **knapsack** problem as it can be interpreted in terms of a hiker who can only carry a certain total weight in his/her knapsack (rucksack). The hiker has a number of items that he/she would like to take, each of which has a particular value. The problem is to decide which items should be packed so that the total value is a maximum, subject to the weight restriction.

This type of problem will be solved using a technique called the **branch and bound method**. How it works will be shown using the following particular problem.

Suppose a traveller wishes to buy some books for his journey. He estimates the time it will take to read each of five books and notes the cost of each one :

Book	A	B	C	D	E
Cost (£)	4	6	3	2	5
Reading time (hours)	5	9	4	4	4

Which of these books should he buy to maximise his total reading time without spending more than £8?

Activity 5

By trial and error, find the solution to the traveller's problem.

As you have probably seen, with just a few items it is easy enough to find the optimum solution. However, in the example above, if there was a choice of, say, 10 or 15 books, the problem of finding the optimum solution would now be far more complex.

The first step in the branch and bound method is to list the items in decreasing order of reading time per unit cost.

Item	A	B	C	D	E	
Cost	4	6	3	2	5	('weight')
Reading time	5	9	4	4	4	('value')
Reading time per unit cost	1.25	1.5	1.33	2	0.8	('value per unit weight')

Reordering,

Number	1	2	3	4	5
Item	D	B	C	A	E
Cost (w)	2	6	3	4	5
Value (v)	4	9	4	5	4
Value per unit cost	2	1.5	1.33	1.25	0.8

A solution **vector** of the form $(x_1, x_2, x_3, x_4, x_5)$ will be used to denote a possible solution where

$$x_i = \begin{cases} 1 & \text{if item } i \text{ is bought} \\ 0 & \text{if item } i \text{ is not bought} \end{cases}$$

So, for example,

$$\mathbf{x} = (0, 0, 1, 0, 1)$$

means buying books C and E, which have total cost £8 and total value $4 + 4 = 8$ hours of reading time.

The method uses a branching method to search for the optimal solution.

For example let us look at the possible branches from $(1, 0, 0, 0, 0)$.

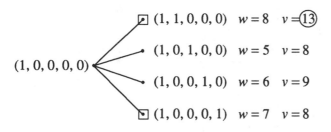

$$(1, 0, 0, 0, 0) \begin{cases} (1, 1, 0, 0, 0) & w = 8 & v = \boxed{13} \\ (1, 0, 1, 0, 0) & w = 5 & v = 8 \\ (1, 0, 0, 1, 0) & w = 6 & v = 9 \\ (1, 0, 0, 0, 1) & w = 7 & v = 8 \end{cases}$$

The 'square' shows that there can be no further branching from this point. For example, there is a square by $(1, 1, 0, 0, 0)$ because the total allowed cost (weight) of 8 has been reached. Also we always add additional 1s to the right of the last 1 so that, for example, you could branch from $(1, 0, 0, 1, 0)$ to $(1, 0, 0, 1, 1)$ but you would never consider branching from $(1, 0, 0, 1, 0)$ to $(1, 0, 1, 1, 0)$. So when, in the example above, x_5 is 1, no further branching is possible : that accounts for the square by $(1, 0, 0, 0, 1)$. Note that the best solution (i.e. maximum v) at this stage is $v = 13$.

You start the full process at the **null** solution $(0, 0, 0, 0, 0)$, now written more simply as $0\,0\,0\,0\,0$, and then keep repeating the process as outlined below.

$$0\,0\,0\,0\,0 \begin{cases} 1\,0\,0\,0\,0 & w = 2 & v = 4 \\ 0\,1\,0\,0\,0 & w = 6 & v = 9 \\ 0\,0\,1\,0\,0 & w = 3 & v = 4 \\ 0\,0\,0\,1\,0 & w = 4 & v = 5 \\ 0\,0\,0\,0\,1 & w = 5 & v = 4 \end{cases}$$

You can now branch from any of the four vertices which are not squared : for example, the branches from $1\,0\,0\,0\,0$ are as shown above and the branches from $0\,1\,0\,0\,0$ are as shown below :

$$0\,1\,0\,0\,0 \begin{cases} 0\,1\,1\,0\,0 & w = 9 \\ 0\,1\,0\,1\,0 & w = 10 \\ 0\,1\,0\,0\,1 & w = 11 \end{cases}$$

Continue in this way to give one single diagram, as shown on the next page :

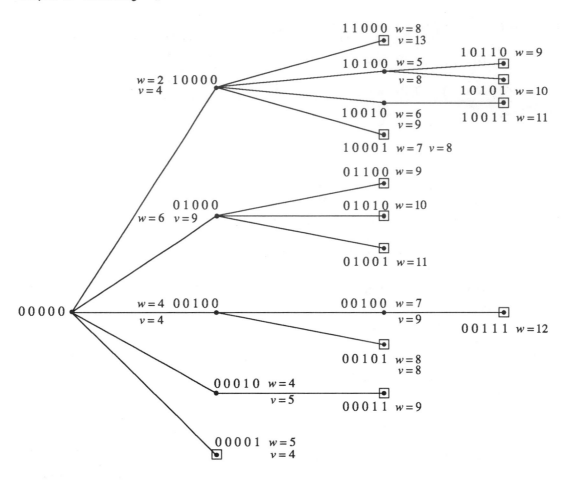

Then from amongst the squared vertices with $w \leq 8$, you find the one with the highest v, namely 1 1 0 0 0 in the this case. That means that the traveller should take books D and B, giving a total reading time of 13 hours.

Exercise 13C

1. Suppose that a hiker can pack up to 9 kg of items and that the following items are available to take. The value of each item is also specified.

Item	A	B	C	D	E
Weight (kg)	3	8	6	4	2
Value	2	12	9	3	5

Use the branch and bound method to find the items that can be taken which give a maximum total value.

2. The manager of a firm which has installed a small computer system has received requests from four potential users. The computer will be run for up to 24 hours per day but can cope with only one user at a time. The manager estimates that the number of hours of computer time required by each user per day, and the consequent likely income for the firm, are as follows:

User	A	B	C	D
Use (hours/day)	8	12	13	4
Income (£1000/year)	72	102	143	38

Use the branch and bound method to determine which users should be allocated time on the computer system so as to maximise the total income.

3. A machine in a factory can be used to make any one of five items, A, B, C, D and E. The time taken to produce each item, and the value of each item, are shown in the table opposite.

If the machine is available for only 10 days, use the branch and bound method to determine which of the items should be produced so that the total value is as large as possible.

Item	A	B	C	D	E
Production time (in days)	3	7	2	4	4
Value	3	14	3	7	8

13.4 Miscellaneous Exercises

1. A project consists of ten activities A-J with the following durations (in hours). There are no precedence relations.

Activity	A	B	C	D	E	F	G	H	I	J
Duration	2	3	4	5	6	7	8	9	10	11

(a) Find the minimum number of workers needed to complete this project in 16 hours.

(b) Use (i) the first-fit packing method,

    (ii) the first-fit decreasing method.

Does either of these methods produce an optimum solution?

2. A hiker wishes to take with her a number of items. Their weights and values are given in the table below.

Item	A	B	C	D	E
Weight (kg)	5	4	7	3	6
Value	3	3	4	2	4

If the maximum weight she can carry is 12 kg, find by trial and error the best combination of items to carry. Use the branch and bound method to confirm your solution as optimal.

3. A small firm orders planks of wood of length 20 m. Each week the firm orders a certain number of planks and then has to meet the orders for that week. Use a bin-filling method to find the minimum number of planks required to meet the following weekly orders.

(a)
Length	Number Required
3 m	5
4 m	6
5 m	2
7 m	2
8 m	1
9 m	1

(b)
Length	Number Required
11 m	1
9 m	1
7 m	2
5 m	2
3 m	12

(c)
Length	Number Required
15 m	1
12 m	2
11 m	1
7 m	3
3 m	2

Does the method always produce the optimum solution?

4. The durations of nine activities A to I are given below, in days:

Activity	A	B	C	D	E	F	G	H	I
Duration	5	9	1	7	3	2	8	4	6

There are no precedence relations, but each activity must be completed by just one worker. We wish to find the minimum number of workers needed to complete all nine activities in 12 days.

(a) What answer is given by the first-fit packing method?

(b) If instead the first-fit decreasing method were used, what answer would then be found?

5. A project consists of eight activities A - H with the following durations (in days). There are no precedence relations.

Activity	A	B	C	D	E	F	G	H
Duration	8	3	9	1	2	6	7	4

It is required to find the minimum number of workers needed to complete this project in 10 days. Each activity is to be completed by a single worker.

(a) What answer is given to this problem by the first-fit packing method?

(b) What answer is given by the first-fit decreasing method?

(In each part you should draw a diagram to show which tasks are allocated to which workers.)

14 DIFFERENCE EQUATIONS I

Objectives

After studying this chapter you should

- be able to detect recursive events within contextual problems;
- be able to recognise and describe associated sequences;
- be able to solve a number of first order difference equations;
- be able to apply solutions of first-order difference equations to contextual problems.

14.0 Introduction

Imagine you are to jump from an aircraft at an altitude of 1000 metres. You want to free-fall for 600 metres, knowing that in successive seconds you fall

$$5, 15, 25, 35, \ ... \ \text{metres.}$$

How many seconds do you count before you pull the rip-cord?

Developing a method for answering this type of question is an aim of this chapter. Perhaps you could attempt the problem now by studying the pattern within the sequence.

The methods employed in this chapter are widely used in applied mathematics, especially in areas such as economics, geography and biology. The above example is physical, but as you will see later there is no need to resort to physical or mechanical principles in order to solve the problem.

Activity 1   Tower of Hanoi

You may be familiar with the puzzle called the **Tower of Hanoi**, in which the object is to transfer a pile of rings from one needle to another, one ring at a time, in as few moves as possible, with never a larger ring sitting upon a smaller one.

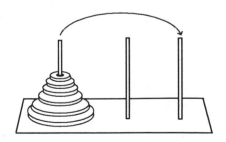

The puzzle comes from the Far East, where in the temple of Benares a priest unceasingly moves a disc each day from an original pile of sixty four discs on one needle to another. When he has finished the world will end!

Try this game for yourself. There may be one in school or you could make one. The puzzle is commonly found among mathematics education software.

Record the number of moves required for initial piles of one, two, three rings, etc. Try to predict the number of moves required for 10 rings and 20 rings. When should the world end?

The solution to the problem involves the idea of recursion (from recur - to repeat). The next section considers a further problem through which the ideas of recursion can be explored. You will meet the Tower of Hanoi again, later on.

14.1 Recursion

Here is a simple sequence linked to a triangular dot pattern. Naturally, these are called the triangle numbers 1, 3, 6, 10, 15, ...

In order to obtain the next term (the sixth), one more row of six dots is added to the fifth term. If a term much further down the sequence were required, you could simply keep adding on 7, then 8, then 9 and so on. This process is called **recursion**.

The process can be described algebraically. Call the first term u_1, the second u_2 and the general term u_n, where n is a positive integer.

So $\qquad u_1 = 1$

$\qquad u_2 = 3$

$\qquad u_3 = 6$

$\qquad$ etc.

In order to find u_n you have to add the number n to u_{n-1}. This gives the expression

$$u_n = u_{n-1} + n$$

Expressions of this type are called **difference equations** (or **recurrence relations**).

What processes in life are recursive? How, if at all, does natural recursion differ from mathematical recursion?

In order to verify that this expression determines the sequence of triangle numbers, the term u_6 is found from using the known value of u_5 :

$$u_5 = 15$$

and

$$u_6 = u_5 + 6$$

$$= 15 + 6$$

$$= 21, \text{ as expected.}$$

Example

If $u_1 = 4$ and $u_n = 2u_{n-1} + 3n - 1$, for $n \geq 2$, find the values of u_2 and u_3.

Solution

$$u_2 = 2u_1 + 3 \times 2 - 1$$

$$= 2 \times 4 + 6 - 1$$

$$= 8 + 6 - 1$$

$$= 13$$

and

$$u_3 = 2u_2 + 3 \times 3 - 1$$

$$= 2 \times 13 + 9 - 1$$

$$= 26 + 9 - 1$$

$$= 34.$$

Activity 2   Dot patterns

Draw a number of dot patterns which increase in a systematic way, for example,

or

For each pattern, write down a difference equation and show that, from knowing u_1, you can use your equation to generate successive terms in your patterns.

What other patterns occur in life? Can they be described using numbers?

Exercise 14A

1. For each equation you are given the first term of a sequence. Find the 4th term in each case :

   (a) $u_1 = 2$ and $u_n = u_{n-1} + 3$, $n \geq 2$

   (b) $u_1 = 1$ and $u_n = 3u_{n-1} + n$, $n \geq 2$

   (c) $u_1 = 0$ and $u_n - u_{n-1} = n + 1$, $n \geq 2$.

2. For each sequence write down a difference equation which describes it:

   (a) 3 5 7 9 11

   (b) 2 5 11 23 47

   (c) 1 2 5 14 41.

3. A vacuum pump removes one third of the remaining air in a cylinder with each stroke. Form an equation to represent this situation. After how many strokes is just 1 / 1 000 000 of the initial air remaining?

4. Write down the first four terms of each of these sequences and the associated difference equation.

   (a) $u_n = \sum_{r=1}^{n} (2r - 1)$ (b) $u_n = \sum_{r=1}^{n} (10 - r)$

   (c) $u_n = \sum_{r=1}^{n} 3(2r + 1)$

5. Write a simple computer program (say in Basic) which calculates successive terms of a sequence from a difference equation you have met. Here is one for the triangle numbers to help you.

```
10  REM"SEQUENCE  OF  TRIANGLE  NUMBERS"
20  INPUT"NUMBER  OF  TERMS  REQUIRED";N
30  CLS
40  U=0:PRINT"FIRST  ";N;"  TRIANGLE
    NUMBERS  ARE"
50  FOR  i=1  TO  N:U=U+i:PRINTU:NEXTi
60  STOP
```

14.2 Iteration

Consider again the Tower of Hanoi.

You should have found a sequence of minimum moves as follows :

Number of rings	1	2	3	4	5	6	...
Number of moves	1	3	7	15	31	63	...

Successive terms are easily found by doubling and adding one to the previous term, but it takes quite a long time to reach the sixty-fourth term, which by the way is about 1.85×10^{19} or 18.5 million million million!

A different approach is to try to work 'backwards' from the nth term u_n, rather than starting at u_1, and building up to it. In this case :

$$u_n = 2u_{n-1} + 1, \quad n \geq 2 \tag{1}$$

For example,

$$u_6 = 2u_5 + 1$$

$$= 2 \times 15 + 1$$

$$= 31.$$

In a similar way, u_{n-1} can be written in terms of u_{n-2} as

$$u_{n-1} = 2u_{n-2} + 1. \qquad\qquad (2)$$

Then equation (2) can be substituted into equation (1) to give

$$u_n = 2(2u_{n-2} + 1) + 1$$

$$= 4u_{n-2} + 2 + 1.$$

Repeating this process using $u_{n-2} = 2u_{n-3} + 1$ gives

$$u_n = 2(4u_{n-3} + 2 + 1) + 1$$

$$= 8u_{n-3} + 4 + 2 + 1$$

$$\Rightarrow \quad u_n = 2^3 u_{n-3} + 2^2 + 2^1 + 2^0.$$

You can see a pattern developing. Continuing until u_n is expressed in terms of u_1, gives

$$u_n = 2^{n-1} u_1 + 2^{n-2} + 2^{n-3} + \ldots + 2^2 + 2^1 + 2^0$$

$$= 2^{n-1} + 2^{n-2} + \ldots + 2^2 + 2^1 + 2^0, \text{ (as } u_1 = 1). \quad (3)$$

You should recognise (3) as a geometric progression (GP) with first term 1 and common ratio 2. Using the formula for the sum to n terms of a GP,

$$u_n = \frac{1(2^n - 1)}{2 - 1} = 2^n - 1, \quad n \geq 1.$$

This is the solution to the differential equation $u_n = 2u_{n-1} + 1$. This process involved the repeated use of a formula and is known as **iteration**.

You can now see how easy it is to calculate a value for u_n.

For example,

$$u_{100} = 2^{100} - 1$$

$$\approx 127 \times 10^{30}.$$

Activity 3

You invest £500 in a building society for a number of years at a rate of 10% interest per annum. Find out how much will be in the bank after 1, 2 or 3 years. Try to write down the difference equation which describes the relationship between the amount in the bank, u_n, at the end of the nth year with the amount u_{n-1} at the end of the previous year. Solve your equation by iteration in the way shown for the Tower of Hanoi problem.

Use your solution to find the amount accrued after 10 years.

How long does it take for your money to double?

Exercise 14B

1. Solve by iteration, giving u_n in terms of u_1 .

   (a) $u_n = u_{n-1} + 2, \ n \ge 2$

   (b) $u_n = 4u_{n-1} - 1, \ n \ge 2$

   (c) $u_n = 3u_{n-1} + 2, \ n \ge 2$

2. A population is increasing at a rate of 25 per thousand per year. Define a difference equation which describes this situation. Solve it and find the population in 20 years' time, assuming the population is now 500 million. How long will it take the population to reach 750 million?

3. Calculate the monthly repayment on a £500 loan over 2 years at an interest rate of $1\frac{1}{2}$ % per month.

14.3 First order difference equations

Equations of the type $u_n = ku_{n-1} + c$, where k, c are constants, are called **first order linear difference equations** with constant coefficients. All of the equations you have met so far in this chapter have been of this type, except for the one associated with the triangle numbers in Section 14.1

For the triangle numbers $u_n = u_{n-1} + n$, and since n is not constant, this is not a linear difference equation with constant coefficients.

If the equation is of the type

$$u_n = ku_{n-1},$$

then the solution can be found quite simply. You know that $u_n = ku_{n-1}$, $u_{n-1} = ku_{n-2}$, and so on, so that the difference equation becomes

$$u_n = ku_{n-1}$$

$$= k\left(ku_{n-2}\right)$$

$$= k^2 u_{n-2}$$

$$= k^2\left(ku_{n-3}\right)$$

$$= k^3 u_{n-3}$$

etc.

$$\Rightarrow \quad u_n = k^{n-1}u_1.$$

So if $\quad u_n = ku_{n-1}, \quad k$ constant, $n \geq 2$,

then its solution is $\quad u_n = k^{n-1}u_1$.

Example

Solve $u_n = 5u_{n-1}$, where $u_1 = 2$, and find u_5.

Solution

$$u_n = 5^{n-1}u_1$$

$$= 2 \times 5^{n-1}$$

$$\Rightarrow \quad u_5 = 2 \times 5^4$$

$$= 1250.$$

If the equation is of the type $u_n = ku_{n-1} + c$ then a general solution can be found as follows :

$$u_n = ku_{n-1} + c$$

$$= k\left(ku_{n-2} + c\right) + c$$

$$= k^2 u_{n-2} + kc + c$$

$$= k^2\left(ku_{n-3} + c\right) + kc + c$$

$$\ldots \qquad \ldots \qquad \ldots$$

$$= k^{n-1}u_1 + k^{n-2}c + k^{n-3}c + \ldots + kc + c$$

$$= k^{n-1}u_1 + c\left(1 + k + k^2 + \ldots + k^{n-2}\right).$$

Again, there is a GP, $1 + k + k^2 + \ldots + k^{n-2}$, to be summed. This has a first term of 1, $n-1$ terms and a common ratio of k. So, provided that $k \neq 1$, the sum is

$$\frac{1\left(k^{n-1}-1\right)}{k-1}.$$

So if the difference equation is of the form $u_n = ku_{n-1} + c$, where k and c are constant, and $k \neq 1$, then it has solution

$$\boxed{u_n = k^{n-1}u_1 + \frac{c\left(k^{n-1}-1\right)}{k-1}}$$

The case $k = 1$ will be dealt with in a moment.

Example

Solve $u_n = 2u_{n-1} - 3$, $n \geq 2$, given $u_1 = 4$.

Solution

Using the formula above,

$$u_n = 2^{n-1} \times 4 - \frac{3\left(2^{n-1}-1\right)}{2-1}$$

$$= 4 \times 2^{n-1} - 3 \times 2^{n-1} + 3$$

$$= 2^{n-1} + 3.$$

You can see this formula works by finding, say, the value of u_2 from the difference equation as well.

Using the difference equation,

$$u_2 = 2u_1 - 3$$

$$= 2 \times 4 - 3$$

$$= 8 - 3$$

$$= 5.$$

Using the formula,

$$u_2 = 2^1 + 3$$

$$= 5.$$

Special case

In the formula for u_n, k cannot equal one. In this case, when $k = 1$, the difference equation is of the type

$$u_n = u_{n-1} + c, \quad n \geq 2.$$

This has the simple solution

$$\boxed{u_n = u_1 + (n-1)c}$$

This is the type of sequence in the parachute jump problem of Section 14.0. If, in that example, you let u_n be the number of metres fallen after n seconds, then

$$u_n = u_{n-1} + 10, \quad u_1 = 5$$

$$\Rightarrow \quad u_n = 10(n-1) + 5$$

For example,

$$u_4 = 10 \times 3 + 5$$

$$= 30 + 5$$

$$= 35.$$

You can now use the formula to solve the original problem of how long it takes to fall 600 metres.

Notation

In some cases, it is convenient to number the terms of a sequence,

$$u_0, u_1, u_2 \ldots$$

rather than

$$u_1, u_2, u_3, \ldots$$

This will often be the case in the next chapter. The different numbering affects both the difference equation and its solution.

For example, look again at the triangle numbers

$$1, 3, 6, 10, 15, \ldots$$

If these are denoted by u_1, u_2, u_3, ..., then, for example, $u_3 = 6$, $u_4 = 10$, $u_4 = u_3 + 4$, and in general $u_n = u_{n-1} + n$. The solution turns out to be

$$u_n = \tfrac{1}{2}n(n+1).$$

If, however, you start again and instead denote that sequence of triangle numbers by u_0, u_1, u_2, ..., then $u_2 = 6$, $u_3 = 10$, $u_3 = u_2 + 4$, in general $u_n = u_{n-1} + (n+1)$. In this case the solution is given by

$$u_n = \tfrac{1}{2}(n+1)(n+2)$$

(which, of course, could be obtained from the previous general solution by replacing n by $n+1$).

If you choose to write sequences beginning with the term u_0, then the solutions to difference equations of the form

$$u_n = ku_{n-1} + c$$

alter somewhat as shown below.

If $\qquad u_n = ku_{n-1} + c, \quad n \geq 1$,

then $\qquad \boxed{u_n = k^n u_0 + \dfrac{c\left(k^n - 1\right)}{k-1}, \quad k \neq 1}$

or, if $u_n = u_{n-1} + c$, then $u_n = u_0 + nc$.

Example

Find the solution of the equation $u_n = 3u_{n-1} + 4$, given $u_0 = 2$.

Solution

$$\begin{aligned}
u_n &= 3^n u_0 + \frac{4\left(3^n - 1\right)}{3-1} \\[2mm]
&= 3^n \times 2 + 2\left(3^n - 1\right) \\[2mm]
&= 2 \times 3^n + 2 \times 3^n - 2 \\[2mm]
&= 4 \times 3^n - 2.
\end{aligned}$$

In the example above the **particular solution** to the difference equation $u_n = 3u_{n-1} + 4$ when $u_0 = 2$ has been found.

If you had not substituted for the value of u_0 (perhaps not knowing u_0) then a **general solution** could have been given as

$$u_n = 3^n u_0 + 2\left(3^n - 1\right).$$

This solution is valid for all sequences which have the same difference equation whatever the initial term u_0.

Example

Find the general solution of the difference equation

$$u_n = u_{n-1} + 4, \quad n \geq 1.$$

Solution

$$u_n = u_0 + 4n.$$

Exercise 14C

1. Write down the general solutions of :

   (a) $u_n = 4u_{n-1} + 2, \quad n \geq 2$

   (b) $u_n = 4u_{n-1} + 2, \quad n \geq 1$

   (c) $u_n = 3u_{n-1} - 5, \quad n \geq 1$

   (d) $u_{n+1} = u_n + 6, \quad n \geq 0$

   (e) $u_n = u_{n-1} - 8, \quad n \geq 2$

   (f) $u_n = -2u_{n-1} + 4, \quad n \geq 1$

   (g) $u_n + 3u_{n-1} - 2 = 0, \quad n \geq 1$

   (h) $u_n + 4u_{n-1} + 3 = 0, \quad n \geq 1$

   (i) $u_n = 4u_{n-1}, \quad n \geq 2$

   (j) $u_{n+1} = 4u_n - 5, \quad n \geq 0$

2. Find the particular solutions of these equations :

   (a) $u_0 = 1$ and $u_n = 3u_{n-1} + 5, \quad n \geq 1$

   (b) $u_1 = 3$ and $u_n = -2u_{n-1} + 6, \quad n \geq 2$

   (c) $u_0 = 4$ and $u_n = u_{n-1} - 3, \quad n \geq 1$

   (d) $u_1 = 0$ and $u_{n+1} = 5u_n + 3, \quad n \geq 1$

   (e) $u_0 = 3$ and $u_{n+1} = u_n + 7, \quad n \geq 0$

   (f) $u_0 = 1$ and $u_n + 3u_{n-1} = 1, \quad n \geq 1$.

14.4 Loans

Activity 4

Find out about the repayments on a loan from a bank, building society or other lending agency. You will need to know the rate of interest per annum, the term of the loan and the frequency of the repayment.

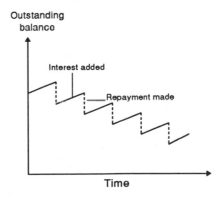

Use these variables in order to set up a difference equation :

$$u_n = \text{amount in £ owing after } n \text{ repayments}$$

$$k = \text{interest multiplier}$$

$$c = \text{repayment in £}$$

It will be of the type described in Section 14.2.

Solve your equation and then evaluate c, the repayment. Remember that there is nothing owing after the final repayment has been made.

How does your result compare with the repayments specified by the lending agency?

The activity is quite difficult. So if you need help an example of a similar type follows.

Example

Find the monthly repayment on a £500 loan at an interest rate of 24% p.a. over 18 months.

Solution

Let u_n be the amount owing after n months, so $u_0 = £500$ and c is the repayment each month. You are given that the interest rate per annum is 24%, so the interest rate per month is assumed to be 2%. (Can you see why this is an approximation?) Then

$$u_n = u_{n-1} + \left(2\% \text{ of } u_{n-1}\right) - c.$$

Here, the interest has been added to the previous outstanding loan, and one month's repayment subtracted.

$$\Rightarrow \quad u_n = u_{n-1} + 0.02 u_{n-1} - c$$

$$u_n = 1.02 u_{n-1} - c.$$

Remembering that you started with u_0, and noting the switch to a negative c, the general solution of this type of equation is

$$u_n = k^n u_0 + \frac{c\left(k^n - 1\right)}{(k-1)}$$

where in this case $k = 1.02$.

So
$$u_n = (1.02)^n u_0 - \frac{c\left(1.02^n - 1\right)}{1.02 - 1}$$

$$= 1.02^n u_0 - 50c\left(1.02^n - 1\right)$$

or
$$u_n = 1.02^n u_0 - 50c\left(1.02^n - 1\right).$$

At the end of the term $n = 18$, so

$$u_{18} = 1.02^{18} u_0 - 50c\left(1.02^{18} - 1\right).$$

But $u_0 = 500$ and $u_{18} = 0$ (the loan has been paid off)

$$\Rightarrow \quad 0 = 500 \times 1.02^{18} - 50c\left(1.02^{18} - 1\right)$$

$$\Rightarrow \quad c = \frac{500 \times 1.02^{18}}{50\left(1.02^{18} - 1\right)}$$

$$\approx \frac{714.123}{21.4123}$$

$$\approx £33.35.$$

Repayments are made at the rate of £33.35 per month.

In the example on the previous page, as in many situations involving loans, the monthly rate is taken as one twelfth of the annual rate. Is the rate of 24% p.a. justifiable as a measure of the interest paid? What is the significance of an APR?

Exercise 14D

1. (a) Find the general solution of
   $$u_n = 3u_{n-1} + 4, n \geq 2$$

   (b) Find the general solution of
   $$u_n = \tfrac{1}{2}u_{n-1} + 2, n \geq 2$$

   (c) Solve $q_n = q_{n-1} + 3$, given $q_3 = 7$, $n \geq 2$

   (d) Solve $a_n - 2a_{n-1} = 0$, given $a_1 = 4$, $n \geq 2$

   (e) Solve $b_n = 4b_{n-1} + 5$, given $b_1 = 2$, $n \geq 2$

2. Form and solve the difference equation associated with the sequence
   $$7, 17, 37, 77, 157 \ldots$$

3. Find the monthly repayment on a £400 loan over a period of $1\tfrac{1}{2}$ years at an interest rate of 15% p.a.

4. Find the monthly repayment on a £2000 loan over a term of 3 years at an interest rate of 21% p.a.

5. Write a computer program to solve the type of problem in Question 3.

6. A steel works is increasing production by 1% per month from a rate of 2000 tonnes per month. Orders (usage of the steel) remain at 1600 tonnes per month. How much steel will be stock-piled after periods of 12 months and 2 years?

7. A loan of £1000 is taken out at an interest rate of 24% p.a. How long would it take to repay the loan at a rate of £50 per month.

   [**Note**: this problem is similar to Question 3, except that the value of n cannot easily be found after you have solved the difference equation. A search technique on your calculator or a simple computer program will be necessary.]

14.5 Non-homogeneous linear equations

So far you have met equations of the form

$$u_n = ku_{n-1}, \quad k \text{ constant}$$

This is called a **homogeneous** equation, involving terms in u_n, and u_{n-1} only.

You have also solved equations of the form

$$u_n = ku_{n-1} + c, \quad k, c \text{ constant}$$

This is a **non-homogeneous** equation, due to the extra term c, but k and c are still constant.

Earlier you met the difference equation associated with the triangle numbers

$$u_n = u_{n-1} + n$$

This is again a non-homogeneous equation, but the term n is not constant.

In general, equations of the type $u_n = ku_{n-1} + f(n)$ are difficult to solve. Expanding u_n as a series is occasionally successful - it depends usually on how complicated $f(n)$ is. In the next chapter you will meet two methods for solving these equations, and also second order equations of a similar type (these involve terms in u_n, u_{n-1} and u_{n-2}). One method is based on trial and error and the other on the use of **generating functions**. Simpler equations such as the one for the triangle numbers can be solved quite quickly by expanding u_n as a series. Thus

$$u_n = (u_{n-2} + n - 1) + n$$

$$= u_{n-2} + (n-1) + n$$

$$= (u_{n-3} + n - 2) + (n-1) + n$$

$$= \dots \quad \dots \quad \dots \quad \dots \quad \dots$$

$$= 1 + 2 + 3 + \dots + (n-1) + n.$$

This is an **arithmetic progression** (AP). Summing gives

$$u_n = \frac{n}{2}(n+1).$$

You can check this result by evaluating, say, u_5.

Activity 5

Try out this method on one of the 'dot' patterns which you designed in Activity 2.

The problems in the following exercise should all yield to a method similar to the one above. The function $f(n)$ will be restricted to the form n, n^2 or k^n, where k is a constant. This will also be true of later difference equations in the next chapter.

Exercise 14E

1. Solve $u_n = u_{n-1} + n$ if $u_1 = 5$.
2. Find the general solution in terms of u_1:

   (a) $u_n = u_{n-1} + n^2$ (b) $u_n = u_{n-1} + 2^n$

   *(c) $u_n = 2u_{n-1} + n$

3. If $u_n = ku_{n-1} + 5$ and $u_1 = 4$, $u_2 = 17$ find the values of k and u_6.

4. The productivity of an orchard of 2000 trees increases by 5% each year due to improved farming techniques. The farmer also plants a further 100 trees per year. Estimate the percentage improvement in productivity during the next 10 years.

14.6 A population problem

Changes in population can often be modelled using difference equations. The underlying problems are similar to some of the financial problems you have met in this section.

What factors influence demographic change?

Activity 6

The birth and death rates in a country are 40 per thousand and 15 per thousand per year respectively. The initial population is 50 million.

Form a difference equation which gives the population at the end of a year in relation to that at the end of the previous year. Solve the equation and estimate the population in 10 years time.

If, due to the high birth rate, emigration takes place at a rate of 10 000 per year, how will this change your results?

Population analysis can be refined by taking into account many more factors than in the above problem. In particular, the population pyramid shows that different age group sizes affect the whole population in different ways.

Suppose the population of a country is split into two age groups:

Group 1   consisting of the 0 - 12 year olds, and

Group 2   consisting of the rest,

and assume that births only occur in Group 2. Each group will have its own death rate.

Define $p_1(t)$ as the population of the 0 - 12 group in year t

$p_2(t)$ as the population of the 13+ group in year t

b as the birth rate

d_1 as the death rate in the 0 - 12 group

d_2 as the death rate in the 13+ group.

One further assumption made is that in each year one twelfth of the survivors from Group 1 progress to Group 2.

For **Group 1**

$$p_1(t+1) \;=\; \underbrace{b\, p_2(t)}_{\substack{\text{those born} \\ \text{to people in} \\ \text{Group 2}}} \;+\; \underbrace{\tfrac{11}{12} p_1(t)\left(1-d_1\right)}_{\substack{\text{of those who survive,} \\ \tfrac{11}{12} \text{ remain in Group 1}}}$$

Remember that $\frac{1}{12}$ of the survivors of Group 1 transfer to Group 2 each year.

For **Group 2**

$$p_2(t+1) = \tfrac{1}{12} p_1(t)(1-d_1) + p_2(t)(1-d_2)$$

Using matrices, both equations can be more simply written as

$$\begin{bmatrix} p_1(t+1) \\ p_2(t+1) \end{bmatrix} = \begin{bmatrix} \tfrac{11}{12}(1-d_1) & b \\ \tfrac{1}{12}(1-d_1) & 1-d_2 \end{bmatrix} \begin{bmatrix} p_1(t) \\ p_2(t) \end{bmatrix}$$

Let $P_t \equiv \begin{bmatrix} p_1(t) \\ p_2(t) \end{bmatrix}$

and $\qquad A = \begin{bmatrix} \frac{11}{12}(1-d_1) & b \\ \frac{1}{12}(1-d_1) & 1-d_2 \end{bmatrix}$

then $\qquad P_{t+1} = AP_t$.

This is a difference equation using matrices! Using the earlier solution to this type of equation you can see that

$$P_t = A^t P_0$$

where P_0 is the initial population .

Suppose that $p_1(0) = 5$ million, $p_2(0) = 15$ million, and that the population parameters are

$$b = 0.4$$

$$d_1 = 0.016$$

$$d_2 = 0.03$$

then the above solution for P_t becomes

$$P_t = \begin{bmatrix} \frac{11}{12}(1-0.016) & 0.04 \\ \frac{1}{12}(1-0.016) & 1-0.03 \end{bmatrix}^t \begin{bmatrix} 5 \\ 15 \end{bmatrix}$$

$$= \begin{bmatrix} 0.902 & 0.04 \\ 0.082 & 0.97 \end{bmatrix}^t \begin{bmatrix} 5 \\ 15 \end{bmatrix}$$

For example, the population after one year can be calculated by simple matrix multiplication :

$$P_1 = \begin{bmatrix} 0.902 & 0.04 \\ 0.082 & 0.97 \end{bmatrix} \begin{bmatrix} 5 \\ 15 \end{bmatrix}$$

$$= \begin{bmatrix} 5.11 \\ 14.96 \end{bmatrix}$$

The total population is therefore 20.07 million.

If the population in 10 years is required, then it is fairly straightforward to evaluate A^{10} using a simple program for matrix multiplication with a computer or a modern graphic/programmable calculator.

Activity 7

Produce a simple program which will multiply matrices as required in the above example. Use it to evaluate the population in 10 years time.

Activity 8

Find a population pyramid with associated birth and death rates for a country of your choice and model the population growth (or decay) as in the above example.

You may wish to consider more than two age groupings. This may prove quite difficult and you will need to adapt your program to multiply larger matrices.

14.7 Miscellaneous Exercises

1. Find the general solution to these difference equations in terms of u_1:

   (a) $u_n = 2u_{n-1}$ (b) $u_n - 3u_{n-1} = 3$

   (c) $u_n - 3u_{n-1} = n$

2. By writing down a difference equation and solving it, find the tenth term of this sequence:

   $$2 \quad 4 \quad 10 \quad 28 \quad 82 \quad 244.$$

3. Find the monthly repayment on a loan of £600 over a period of 12 months at a rate of interest of 3% per month.

4. The population of a country is $12\frac{1}{2}$ million. The birth rate is 0.04, the death rate is 0.03 and 50 000 immigrants arrive in the country each year. Estimate the population in 20 years' time.

5. In a round robin tournament every person (or team) plays each of the others. If there are n players, how many more games are needed if one more player is included? Use this result to set up a difference equation, solve it, and then evaluate the number of games needed for 20 players (this confirms the answer found to Question 4(b) in Exercise 6A).

6. If $u_n = pu_{n-1} + q$, $n \geq 1$, and $u_1 = 2$, $u_2 = 3$, $u_3 = 7$, find the value of u_6.

7. Compare the monthly repayment on a mortgage of £30 000 at an interest rate of 12% p.a. over 25 years for the two standard types of mortgage :

   (a) a **repayment** mortgage, where you pay a fixed amount each month, the interest is calculated each month on the remaining debt, and the amount of repayment is calculated so that the debt is paid off after 25 years;

   (b) an **endowment** mortgage where the loan stays at £30 000 for the whole of the 25 years, you pay the interest on that and an additional £40 a month for a type of insurance policy known as an endowment policy. At the end of the 25 years the insurance policy matures and the insurance firm pays off the debt.

   (You will find that the latter costs more, but in practice when the insurance policy matures it pays well over the £30 000, thus giving you an additional lump sum back.)

8. A population of 100 million can be divided into age groups. **Group 1**, 0-16 years, has a death rate of 0.025 (no birth rate) and a population of 60 million. **Group 2**, 17+ years, has a birth rate of 0.04 and a death rate of 0.03. Investigate the growth/decay of the population. Make a prediction for the population size in 3 years' time.

9. A person is repaying a loan of £5000 at £200 per month. The interest rate is 3% per month. How long will it take to repay the loan?

10. Within a population of wild animals the birth rate is 0.2, while the death rate is 0.4. Zoos worldwide are attempting to reintroduce animals. At present the population is 5000 and 100 animals per year are being introduced. The rate of increase of introduction is 20% per year. Will the animals survive in the long run?

11. Investigate the problem of the Tower of Hanoi with four needles for the rings.

12. Use iteration to solve the recurrence relation

$$u_n = \frac{u_{n-1}}{(n+2)}, \; n \geq 2$$

subject to the initial condition $u_1 = \frac{1}{6}$.

15 DIFFERENCE EQUATIONS 2

Objectives

After studying this chapter you should

- be able to obtain the solution of any linear homogeneous second order difference equation;

- be able to apply the method of solution to contextual problems;

- be able to use generating functions to solve non-homogeneous equations.

15.0 Introduction

In order to tackle this chapter you should have studied a substantial part of the previous chapter on first order difference equations. The problems here deal with rather more sophisticated equations, called second order difference equations, which derive from a number of familiar contexts. This is where the rabbits come in.

It is well known that rabbits breed fast. Suppose that you start with one new-born pair of rabbits and every month any pair of rabbits gives birth to a new pair, which itself becomes productive after a period of two months. How many rabbits will there be after n months? The table shows the results for the first few months.

Month	1	2	3	4	5	6
No. of pairs (u_n)	1	1	2	3	5	8

The sequence u_n is a famous one attributed to a 13th century mathematician *Leonardo Fibonacci* (c. 1170-1250). As you can see the next term can be found by adding together the previous two.

The nth term u_n can be written as

$$u_n = u_{n-1} + u_{n-2}$$

and difference equations like this with terms in u_n, u_{n-1} and u_{n-2} are said to be of the **second order** (since the difference between n and $n-2$ is 2).

Activity 1   Fibonacci numbers

The **Fibonacci numbers** have some remarkable properties. If you divide successive terms by the previous term you obtain the sequence,

$$\frac{1}{1}, \frac{2}{1}, \frac{3}{2}, \frac{5}{3}, \ldots = 1, \ 2, \ 1.5, \ 1.\dot{6}, \ldots$$

Continue this sequence, say to the 20th term, and find its reciprocal. What do you notice? Can you find an equation with a solution which gives you the limit of this sequence?

Exercise 15A

1 If $u_n = 2u_{n-1} + u_{n-2}$ and $u_1 = 2$, $u_2 = 5$, find the values of u_3, u_4, u_5.

2. $u_n = pu_{n-1} + qu_{n-2}$ describes the sequence 1, 2, 8, 20, 68, ... Find p and q.

3. If F_n is a term of the Fibonacci sequence, investigate the value of $F_{n+1} F_{n-1} - F_n^2$.

4. What sequences correspond to the difference equation $u_n = u_{n-1} - u_{n-2}$, $n \geq 3$? Choose your own values for u_1 and u_2.

5. Find the ratio of the length of a diagonal to a side of a regular pentagon. What do you notice?

6. Investigate the limit of $\dfrac{u_n}{u_{n-1}}$ if $u_n = u_{n-1} + 2u_{n-2}$.

7. Show that $F_1 + F_2 + F_3 + \ldots + F_n = F_{n+2} - 1$, where F_n is a term of the Fibonacci sequence.

15.1 General solutions

When you solved difference equations in the previous chapter, any **general** solution had an unknown constant left in the solution. Usually this was u_1 or u_0.

Example

Solve $u_n = 4u_{n-1} - 3$.

Solution

$$u_n = 4^n u_0 - \frac{3\left(4^n - 1\right)}{3}$$

$$= 4^n u_0 - \left(4^n - 1\right)$$

$$= 4^n \left(u_0 - 1\right) + 1$$

Alternatively you could write

$$u_n = A4^n + 1, \text{ replacing } u_0 - 1 \text{ by A.}$$

This first order equation has one arbitrary constant in its general solution. Knowing the value of u_0 would give you a **particular solution** to the equation.

Say $u_0 = 4$ then

$$4 = A.4^0 + 1 = A + 1.$$

So $A = 3$ and

$$u_n = 3 \times 4^n + 1.$$

In a similar way, **general solutions** to second order equations have two arbitrary constants. Unfortunately, an iterative technique does not work well for these equations, but, as you will see, a guess at the solution being of a similar type to that for first order equations does work.

Suppose

$$\boxed{u_n = pu_{n-1} + qu_{n-2}} \qquad \qquad (1)$$

where p, q are constants, $n \geq 2$.

This is a **second order homogeneous linear difference equation** with constant coefficients.

As a solution, try $u_n = Am^n$, where m and A are constants. This choice has been made because $u_n = k^n u_0$ was the solution to the first order equation $u_n = ku_{n-1}$.

Substituting $u_n = Am^n$, $u_{n-1} = Am^{n-1}$ and $u_{n-2} = Am^{n-2}$ into equation (1) gives

$$Am^n = Apm^{n-1} + Aqm^{n-2}$$

$$\Rightarrow \quad Am^{n-2}\left(m^2 - pm - q\right) = 0.$$

If $m = 0$, or $A = 0$, then equation (1) has trivial solutions (i.e. $u_n = 0$). Otherwise, if $m \neq 0$ and $A \neq 0$ then

$$m^2 - pm - q = 0.$$

This is called the **auxiliary equation** of equation (1).

It has the solution

$$m_1 = \frac{p + \sqrt{p^2 + 4q}}{2} \quad \text{or} \quad m_2 = \frac{p - \sqrt{p^2 + 4q}}{2}.$$

m_1 and m_2 can be real or complex. The case where $m_1 = m_2$ is special, as you will see later.

Suppose for now that $m_1 \neq m_2$, then it has been shown that both $u_n = Am_1^n$ and $u_n = Bm_2^n$ are solutions of (1), where A and B are constants.

Can you suggest the form of the general solution?

In fact it is easy to show that a **linear** combination of the two solutions is also a solution. This follows since both Am_1^n and Bm_2^n satisfy equation (1) giving

$$Am_1^n = Am_1^{n-1}p + Am_1^{n-2}q$$

and

$$Bm_2^n = Bm_2^{n-1}p + Bm_2^{n-2}q$$

$$\Rightarrow \quad Am_1^n + Bm_2^n = p\left(Am_1^{n-1} + Bm_2^{n-1}\right) + q\left(Am_1^{n-2} + Bm_2^{n-2}\right).$$

So $Am_1^n + Bm_2^n$ is also a solution of equation (1), and can in fact be shown to be the general solution. That is **any** solution of (1) will be of this form.

In summary, the general solution of $u_n = pu_{n-1} + qu_{n-2}$ is

$$\boxed{u_n = Am_1^n + Bm_2^n, \quad m_1 \neq m_2}$$

where A, B are arbitrary constants and m_1, m_2 are the solutions of the auxiliary equation $m^2 - pm - q = 0$.

Example

Find the general solution of $u_n = 2u_{n-1} + 8u_{n-2}$.

Solution

The auxiliary equation is $m^2 - 2m - 8 = 0$.

This has solutions $m_1 = 4$ and $m_2 = -2$.

The general solution is therefore

$$u_n = A4^n + B(-2)^n.$$

Example

Solve $u_n + 3u_{n-2} = 0$, $n \geq 3$, given that $u_1 = 1$ and $u_2 = 3$.

Solution

The auxiliary equation is $m^2 + 3 = 0$.

$$\Rightarrow \quad m^2 = -3$$

$$\Rightarrow \quad m_1 = \sqrt{3}\,i \quad \text{and} \quad m_2 = -\sqrt{3}\,i \quad \text{(where } i = \sqrt{-1}\text{)}.$$

The **general solution** to the equation is therefore

$$u_n = A\left(\sqrt{3}\,i\right)^n + B\left(-\sqrt{3}\,i\right)^n.$$

When $n = 1$, $u_1 = 1$ and since $u_1 = A\left(\sqrt{3}\,i\right)^1 + B\left(-\sqrt{3}\,i\right)^1$

$$\Rightarrow \quad 1 = A\sqrt{3}\,i - B\sqrt{3}\,i$$

$$\frac{1}{\sqrt{3}\,i} = A - B. \tag{2}$$

When $n = 2$, $u_2 = 3$ and $u_2 = A\left(\sqrt{3}\,i\right)^2 + B\left(-\sqrt{3}\,i\right)^2$

$$\Rightarrow \quad 3 = -A3 - B3$$

$$\Rightarrow \quad -1 = A + B. \tag{3}$$

Adding (2) and (3) gives

$$2A = -1 + \frac{1}{\sqrt{3}\,i} = -1 - \frac{i}{\sqrt{3}}$$

$$\Rightarrow \quad A = -\frac{1}{2}\left(1 + \frac{i}{\sqrt{3}}\right)$$

and

$$B = -\frac{1}{2}\left(1 - \frac{i}{\sqrt{3}}\right).$$

Thus the particular solution to the equation, for $u_1 = 1$, $u_2 = 3$, is given by

$$u_n = -\frac{1}{2}\left(1 + \frac{i}{\sqrt{3}}\right)\left(\sqrt{3}\,i\right)^n - \frac{1}{2}\left(1 - \frac{i}{\sqrt{3}}\right)\left(-\sqrt{3}\,i\right)^n \tag{4}$$

Although this solution is given in terms of the complex number $i = \sqrt{-1}$, it is in fact always a real number.

Activity 2

Show that equation (4) gives $u_1 = 1$ and $u_2 = 3$. Also use this equation to evaluate u_3 and u_4, and check these answers directly from the original difference equation, $u_n + 3u_{n-2} = 0$.

Exercise 15B

1. Find the general solutions to

    (a) $u_n = u_{n-1} + 6u_{n-2}$

    (b) $u_n = 4u_{n-1} + u_{n-2}$

    (c) $u_n - u_{n-1} - 2u_{n-2} = 0$.

2. Find the general solution of the difference equation associated with the Fibonacci sequence. Use $u_0 = 1$, $u_1 = 1$, to find the particular solution.

3. Solve $u_n + 4u_{n-2} = 0$, $n \geq 3$, if $u_1 = 2$, $u_2 = -4$.

4. Solve $u_n - 6u_{n-1} + 8u_{n-2} = 0$, $n \geq 3$, given $u_1 = 10$, $u_2 = 28$. Evaluate u_6.

5. Find the nth term of the sequence

    $-3, 21, 3, 129, 147 \ldots$

15.2 Equations with equal roots

When $m_1 = m_2$, the solution in Section 15.1 would imply that

$$u_n = Am_1^n + Bm_1^n$$

$$= m_1^n(A+B)$$

$$= m_1^n C, \quad \text{where} \quad C = A + B.$$

In this case there is really only one constant, compared with the two expected. Trials show that another possibility for a solution to

$$u_n = pu_{n-1} + qu_{n-2} \tag{1}$$

is $u_n = Dnm_1^n$, and as you will see below, this solution, combined with one of the form Cm_1^n, gives a general solution to the equation when $m_1 = m_2$.

If $u_n = Dnm_1^n$ then

$$u_{n-1} = D(n-1)m_1^{n-1}$$

and

$$u_{n-2} = D(n-2)m_1^{n-2}.$$

If $u_n = Dnm_1^n$ is a solution of (1), then $u_n - pu_{n-1} - qu_{n-2}$ should equal zero.

$$u_n - pu_{n-1} - qu_{n-2}$$

$$= Dnm_1^n - pD(n-1)m_1^{n-1} - qD(n-2)m_1^{n-2}$$

$$= Dm_1^{n-2}\left[nm_1^2 - (n-1)pm_1 - (n-2)q\right]$$

$$= Dm_1^{n-2}\left[n\left(m_1^2 - pm_1 - q\right) + pm_1 + 2q\right]$$

$$= Dm_1^{n-2}\left(pm_1 + 2q\right)$$

because $m_1^2 - pm_1 - q = 0$.

Now, the auxiliary equation has equal roots, which means that

$$p^2 + 4q = 0 \quad \text{and} \quad m_1 = \frac{p}{2}.$$

Therefore $\quad u_n - pu_{n-1} - qu_{n-2} = Dm_1^{n-2}\left(p \times \frac{p}{2} + 2q\right)$

$$= 2Dm_1^{n-2}\left(p^2 + 4q\right)$$

$$= 0, \text{ since } p^2 + 4q = 0.$$

So $u_n = Dnm_1^n$ is a solution and therefore $Dnm_1^n + Cm_1^n$ will be also. This can be shown by using the same technique as for the case when $m_1 \neq m_2$.

In summary, when $p^2 + 4q = 0$ the general solution of $u_n = pu_{n-1} + qu_{n-2}$ is

$$\boxed{u_n = Cm_1^n + Dnm_1^n}$$

where C and D are arbitrary constants.

Example

Solve $u_n + 4u_{n-1} + 4u_{n-2} = 0$, $n \geq 3$, if $u_1 = -2$ and $u_2 = 12$. Evaluate u_5.

Solution

The auxiliary equation is

$$m^2 + 4m + 4 = 0$$

$$\Rightarrow \quad (m+2)^2 = 0$$

$$\Rightarrow \quad m_1 = m_2 = -2.$$

Therefore the general solution is

$$u_n = Dn(-2)^n + C(-2)^n$$

or $\qquad u_n = (-2)^n(C + Dn).$

If $\qquad u_1 = -2, \quad -2(C+D) = -2 \qquad \Rightarrow \qquad C+D = 1.$

Also, as $\quad u_2 = 12, \quad 4(C+2D) = 12 \qquad \Rightarrow \qquad C+2D = 3.$

These simultaneous equations can be solved to give $C = -1$ and $D = 2$.

Thus $\qquad u_n = (-2)^n(2n-1)$

and $\qquad u_5 = (-2)^5(10-1)$

$$= -32 \times 9$$

$$= -288.$$

Activity 3

Suppose that a pair of mice can produce two pairs of offspring every month and that mice can reproduce two months after birth. A breeder begins with a pair of new-born mice. Investigate the number of mice he can expect to have in successive months. You will have to assume no mice die and pairs are always one female and one male!

If a breeder begins with ten pairs of mice, how many can he expect to have bred in a year?

Exercise 15C

1. Find the general solutions of
   (a) $u_n - 4u_{n-1} + 4u_{n-2} = 0$
   (b) $u_n = 2u_{n-1} - u_{n-2}.$

2. Find the particular solution of
   $u_n - 6u_{n-1} + 9u_{n-2} = 0$, $n \geq 3$, when $u_1 = 9$, $u_2 = 36$.

3. If $u_1 = 0, u_2 = -4$, solve $u_{n+2} + u_n = 0$, $n \geq 1$, giving u_n in terms of i.

4. Find the particular solution of
   $u_{n+2} + 2u_{n+1} + u_n = 0$, $n \geq 1$, when $u_1 = -1$, $u_2 = -2$.

5. Find the solutions of these difference equations.
   (a) $u_n - 2u_{n-1} - 15u_{n-2} = 0$, $n \geq 3$, given $u_1 = 1$ and $u_2 = 77$.
   (b) $u_n = 3u_{n-2}$, $n \geq 3$, given $u_1 = 0$, $u_2 = 3$.
   (c) $u_n - 6u_{n-1} + 9u_{n-2} = 0$, $n \geq 3$, given $u_1 = 9$, $u_2 = 45$.

6. Form and solve the difference equation defined by the sequence in which the nth term is formed by adding the previous two terms and then doubling the result, and in which the first two terms are both one.

15.3 A model of the economy

In good times, increased national income will promote increased spending and investment.

If you assume that government expenditure is constant (G) then the remaining spending can be assumed to be composed of investment (I) and private spending on consumables (P). So you can model the national income (N) by the equation

$$N_t = I_t + P_t + G, \text{ where } t \text{ is the year number} \qquad (1)$$

If income increases from year $t-1$ to year t, then you would assume that private spending will increase in year t proportionately. So you can write :

$$P_t = AN_{t-1}, \quad \text{where } A \text{ is a constant.}$$

Also, extra private spending should promote additional investment. So you can write :

$$I_t = B(P_t - P_{t+1}), \quad \text{where } B \text{ is a constant.}$$

Substituting for P_t and I_t in (1) gives

$$N_t = AN_{t-1} + B(P_t - P_{t-1}) + G$$

$$= AN_{t-1} + B(AN_{t-1} - AN_{t-2}) + G$$

$$N_t = A(B+1)N_{t-1} - ABN_{t-2} + G. \qquad (2)$$

So far you have not met equations of this type in this chapter. It is a second order difference equation, but it has an extra constant G.

Before you try the activity below, discuss the effects you think the values of A and B will have on the value of N as t increases.

Activity 4

Take, as an example, an economy in which for year 1, $N_1 = 2$ and for year 2, $N_2 = 4$. Suppose that $G = 1$. By using the difference equation (2) above, investigate the change in the size of N over a number of years for different values of A and B.

At this stage you should not attempt an algebraic solution!

Equation (2) is an example of a **non-homogeneous difference equation**. **Homogeneous** second order equations have the form

$$u_n + au_{n-1} + bu_{n-2} = 0$$

There are no other terms, unlike equation (2) which has an additional constant G.

15.4 Non-homogeneous equations

You have seen how to solve homogeneous second order difference equations; i.e. ones of the form given below but where the right-hand side is zero. Turning to non-homogeneous equations of the form

$$u_n + au_{n-1} + bu_{n-2} = f(n)$$

where f is a function of n, consider as a first example the equation

$$6u_n - 5u_{n-1} + u_{n-2} = n, \quad (n \geq 3) \tag{1}$$

Activity 5

Use a computer or calculator to investigate the sequence u_n defined by

$$6u_n - 5u_{n-1} + u_{n-2} = n, \quad (n \geq 3)$$

for different starting values. Start, for example, with $u_1 = 1$, $u_2 = 2$ and then vary either or both of u_1 and u_2. How does the sequence behave when n is large?

From the previous activity, you may have had a feel for the behaviour or structure of the solution. Although its proof is beyond the scope of this text, the result can be expressed as

$$u_n = \begin{pmatrix} \text{general solution of} \\ \text{associated homogeneous} \\ \text{equation} \end{pmatrix} + \begin{pmatrix} \text{one particular} \\ \text{solution of the} \\ \text{full equation} \end{pmatrix}$$

That is, to solve

$$6u_n - 5u_{n-1} + u_{n-2} = n, \quad (n \geq 2)$$

you first find the general solution of the associated homogeneous equation

$$6u_n - 5u_{n-1} + u_{n-2} = 0 \qquad\qquad (2)$$

and, to this, add one particular solution of the full equation.

You have already seen in Section 15.2 how to solve equation (2). The auxiliary equation is

$$6m^2 - 5m + 1 = 0$$

$$\Rightarrow \quad (3m-1)(2m-1) = 0$$

$$\Rightarrow \quad m = \tfrac{1}{3} \text{ or } \tfrac{1}{2}.$$

So the general solution of (2) is given by

$$u_n = A\left(\tfrac{1}{3}\right)^n + B\left(\tfrac{1}{2}\right)^n, \qquad\qquad (3)$$

where A and B are constants.

The next stage is to find one particular solution of the full equation (1).

Can you think what type of solution will satisfy the full equation?

In fact, once you have gained experience in solving equations of this type, you will recognise that u_n will be of the form

$$u_n = a + bn$$

(which is a generalisation of the function on the right-hand side, namely n).

So if $\qquad u_n = a + bn$

$$\Rightarrow \quad u_{n-1} = a + b(n-1)$$

$$\Rightarrow \quad u_{n-2} = a + b(n-2)$$

and to satisfy (1), we need

$$6(a+bn) - 5\big(a+b(n-1)\big) + a + b(n-2) = n$$

$$6a + 6bn - 5a - 5bn + 5b + a + bn - 2b = n$$

$$2a + 3b + n(2b) = n.$$

Each side of this equation is a polynomial of degree 1 in n.

How can both sides be equal?

To ensure that it is satisfied for all values of n, equate cooefficients on each side of the equation.

$$\text{constant term} \quad \Rightarrow \quad 2a + 3b = 0$$

$$n \text{ term} \quad \Rightarrow \quad 2b = 1.$$

So $b = \frac{1}{2}$ and $a = -\frac{3}{4}$, and you have shown that one particular solution is given by

$$u_n = -\frac{3}{4} + \frac{1}{2}n. \tag{4}$$

To complete the general solution, add (4) to (3) to give

$$u_n = A\left(\tfrac{1}{3}\right)^n + B\left(\tfrac{1}{2}\right)^n - \frac{3}{4} + \frac{1}{2}n.$$

Activity 6

Find the solution to equation (1) which satisfies $u_1 = 1$, $u_2 = 2$.

The main difficulty of this method is that you have to 'guess' the form of the particular solution. The table below gives the usual form of the solution for various functions $f(n)$.

$f(n)$	Form of particular solutions
constant	a
n	$a + bn$
n^2	$a + bn + cn^2$
k^n	ak^n (or ank^n in special cases)

The next three examples illustrate the use of this table.

Example

Find the general solution of $6u_n - 5u_{n-1} + u_{n-2} = 2$.

Solution

From earlier work, the form of the general solution is

$$u_n = A\left(\tfrac{1}{3}\right)^n + B\left(\tfrac{1}{2}\right)^n + \begin{pmatrix} \text{one particular} \\ \text{solution} \end{pmatrix}$$

For the particular solution, try

$$u_n = a \quad \Rightarrow \quad u_{n-1} = a \quad \text{and} \quad u_{n-2} = a$$

which on subtituting in the equation gives

$$6a - 5a + a = 2 \implies a = 1.$$

Hence $u_n = 1$ is a particular solution and the general solution is given by

$$u_n = A\left(\tfrac{1}{3}\right)^n + B\left(\tfrac{1}{2}\right)^n + 1.$$

Example

Find the general solution of $6u_n - 5u_{n-1} + u_{n-2} = 2^n$.

Solution

For the particular solution try $u_n = a2^n$, so that $u_{n-1} = a2^{n-1}$, and substituting in the equation gives

$$6a2^n - 5a2^{n-1} + a2^{n-2} = 2^n$$

$$2^{n-2}(6a \times 4 - 5a \times 2 + a) = 2^n$$

$$24a - 10a + a = 4$$

$$\implies \quad a = \tfrac{4}{15}$$

and so the general solution is given by

$$u_n = A\left(\tfrac{1}{3}\right)^n + B\left(\tfrac{1}{2}\right)^n + \left(\tfrac{4}{15}\right)2^n.$$

In the next example, the equation is

$$6u_n - 5u_{n-1} + u_{n-2} = \left(\tfrac{1}{2}\right)^n.$$

Can you see why the usual trial for a particular solution, namely
$u_n = a\left(\tfrac{1}{2}\right)^n$ **will not work?**

Example

Find the general solution of $6u_n - 5u_{n-1} + u_{n-2} = \left(\tfrac{1}{2}\right)^n$.

Solution

If you try $u_n = a\left(\tfrac{1}{2}\right)^n$ for a particular solution, you will not be able to find a value for the constant a to give a solution. This is because the term $B\left(\tfrac{1}{2}\right)^n$ is already in the solution of the associated homogeneous equation. In this special case, try

$$u_n = an\left(\tfrac{1}{2}\right)^n$$

so that $u_{n-1} = a(n-1)\left(\tfrac{1}{2}\right)^{n-1}$

and $u_{n-2} = a(n-2)\left(\tfrac{1}{2}\right)^{n-2}.$

Substituting in the equation gives

$$6an\left(\tfrac{1}{2}\right)^n - 5a(n-1)\left(\tfrac{1}{2}\right)^{n-1} + a(n-2)\left(\tfrac{1}{2}\right)^{n-2} = \left(\tfrac{1}{2}\right)^n$$

$$\left(\tfrac{1}{2}\right)^{n-2}\left(6an\left(\tfrac{1}{2}\right)^2 - 5a(n-1)\tfrac{1}{2} + a(n-2)\right) = \left(\tfrac{1}{2}\right)^n$$

$$\tfrac{3}{2}an - \tfrac{5}{2}an + \tfrac{5}{2}a + an - 2a = \tfrac{1}{4}$$

$$\tfrac{1}{2}a = \tfrac{1}{4} \quad \text{(the } n \text{ terms cancel out).}$$

Hence $a = \tfrac{1}{2}$, and the particular solution is

$$u_n = \tfrac{1}{2}n\left(\tfrac{1}{2}\right)^n = n\left(\tfrac{1}{2}\right)^{n+1}.$$

The general solution is given by

$$u_n = A\left(\tfrac{1}{3}\right)^n + B\left(\tfrac{1}{2}\right)^n + n\left(\tfrac{1}{2}\right)^{n+1}.$$

The examples above illustrate that, although the algebra can become quite complex, the real problem lies in the intelligent choice of the form of the solution. Note that if the usual form does not work, then the degree of the polynomial being tried should be increased by one. The next two sections will illustrate a more general method, not dependent on inspired guesswork!

Exercise 15D

1. Find the general solution of the difference equation $u_n - 5u_{n-1} + 6u_{n-2} = f(n)$ when

   (a) $f(n) = 2$ (b) $f(n) = n$

   (c) $f(n) = 1+n^2$ (d) $f(n) = 5^n$

   (e) $f(n) = 2^n$.

2. Find the complete solution of

   $$u_n - 7u_{n-1} + 12u_{n-2} = 2^n$$

   when $u_1 = 1$ and $u_2 = 1$.

3. Find the general solution of

   $$u_n + 3u_{n-1} - 10u_{n-2} = 2^n$$

   and determine the solution which satisfies $u_1 = 2$, $u_2 = 1$.

15.5 Generating functions

This section will introduce a new way of solving difference equations by first applying the method to homogeneous equations.

A different way of looking at a sequence u_0, u_1, u_2, u_3 is as the coefficients of a power series

$$G(x) = u_0 + u_1 x + u_2 x^2 + \ldots$$

Notice that the sequence and series begin with u_0 rather than u_1. This makes the power of x and the suffix of u the same, and will help in the long run.

$G(x)$ is called the **generating function** for the sequence $u_0, u_1, u_2, \ldots$ This function can be utilised to solve difference equations. Here is an example of a type you have already met, to see how the method works.

Example

Solve $u_n = 3u_{n-1} - 2u_{n-2} = 0$, $n \geq 2$, given $u_0 = 2$, $u_1 = 3$.

Solution

Let
$$G(x) = u_0 + u_1 x + u_2 x^2 + \ldots$$

$$= 2 + 3x + u_2 x^2 + \ldots \qquad (1)$$

Now from the original difference equation

$$u_2 = 3u_1 - 2u_0$$

$$u_3 = 3u_2 - 2u_1$$

$$u_4 = 3u_3 - 2u_2, \quad \text{etc.}$$

Substituting for $u_2, u_3, u_4, \ldots$ into equation (1) gives

$$G(x) = 2 + 3x + (3u_1 - 2u_0)x^2 + (3u_2 - 2u_1)x^3 + \ldots$$

$$= 2 + 3x + \left(3u_1 x^2 + 3u_2 x^3 + 3u_3 x^4 + \ldots\right)$$
$$- \left(2u_0 x^2 + 2u_1 x^3 + 2u_2 x^4 + \ldots\right)$$

$$= 2 + 3x + 3x\left(u_1 x + u_2 x^2 + u_3 x^3 + \ldots\right)$$
$$- 2x^2\left(u_0 + u_1 x + u_2 x^2 + \ldots\right)$$

$$= 2+3x+3x\big(G(x)-u_0\big)-2x^2G(x)$$

$$= 2+3x+3x\big(G(x)-2\big)-2x^2G(x)$$

$$G(x) = 2-3x+3xG(x)-2x^2G(x).$$

Rearranging so that $G(x)$ is the subject gives

$$G(x) = \frac{2-3x}{1-3x+2x^2}.$$

Note that $1-3x+2x^2$ is similar to the auxiliary equation you met previously but **not** the same.

Factorising the denominator gives

$$G(x) = \frac{2-3x}{(1-2x)(1-x)}. \qquad (2)$$

You now use partial fractions in order to write $G(x)$ as a sum of two fractions. There are a number of ways of doing this which you should have met in your pure mathematics core studies.

So $$G(x) = \frac{1}{1-2x}+\frac{1}{1-x}.$$

Now both parts of $G(x)$ can be expanded using the binomial theorem

$$(1-2x)^{-1} = \big(1+2x+(2x)^2+(2x)^3+...\big)$$

and $$(1-x)^{-1} = \big(1+x+x^2+x^3+...\big).$$

This gives $$G(x) = \big(1+2x+(2x)^2+...\big)+\big(1+x+x^2+...\big)$$

$$= (1+1)+(2x+x)+\big(2^2x^2+x^2\big)$$

$$+\big(2^3x^3+x^x\big)+...$$

$$= 2+3x+\big(2^2+1\big)x^2+\big(2^3+1\big)x^3+...$$

As you can see, the nth term of $G(x)$ is $\big(2^n+1\big)x^n$ and the coefficient of x^n is simply u_n - the solution to the difference equation.

So $$u_n = 2^n+1.$$

*Exercise 15E

1. Find the generating function associated with these difference equations and sequences.

   (a) $u_n = 2u_{n-1} + 8u_{n-2}$, given $u_0 = 0$, $u_1 = 1$, $n \geq 2$.

   (b) $u_n + u_{n-1} - 3u_{n-2} = 0$, given $u_0 = 2$, $u_1 = 5$, $n \geq 2$.

   (c) $u_n = 4u_{n-2}$, given $u_0 = 1$, $u_1 = 3$, $n \geq 2$.

   (d) 1, 2, 4, 8, 16, ...

2. Write these expressions as partial fractions :

   (a) $\dfrac{3x-5}{(x-3)(x+1)}$ (b) $\dfrac{1}{(2x-5)(x-2)}$

   (c) $\dfrac{x+21}{x^2-9}$

3. Write these expressions as power series in x, giving the nth term of each series :

   (a) $\dfrac{1}{1-x}$ (b) $\dfrac{1}{1-2x}$ (c) $\dfrac{1}{1+3x}$

   (d) $\dfrac{1}{(1-x)^2}$ (e) $\dfrac{3}{(1+2x)^2}$

4. Solve these difference equations by using generating functions :

   (a) $u_n - 3u_{n-1} + 4u_{n-2} = 0$, given $u_0 = 0$, $u_1 = 20$, $n \geq 2$.

   (b) $u_n = 4u_{n-1}$, given $u_0 = 3$, $n \geq 1$.

5. Find the generating function of the Fibonacci sequence.

6. Find the particular solution of the difference equation $u_{n+2} = 9u_n$, given $u_0 = 5$, $u_1 = -3$, $n \geq 0$.

15.6 Extending the method

This final section shows how to solve non-homogeneous equations of the form :

$$\boxed{u_n + au_{n-1} + bu_{n-2} = f(n)} \quad (a, b \text{ constants}) \qquad (1)$$

using the generating function method. The techniques which follow will also work for first order equations (where $b = 0$).

Example

Solve $u_n + u_{n-1} - 6u_{n-2} = n$, $n \geq 2$, given $u_0 = 0$, $u_1 = 2$.

Solution

Let the generating function for the equation be

$$G(x) = u_0 + u_1 x + u_2 x^2 + \dots.$$

Now work out $\left(1 + x - 6x^2\right)G(x)$.

The term $\left(1 + x - 6x^2\right)$ comes from the coefficients of u_n, u_{n-1} and u_{n-2} in the equation.

Now $\left(1+x-6x^2\right)G(x) = \left(1+x-6x^2\right)\left(u_0 + u_1 x + u_2 x^2 + ...\right)$

$$= u_0 + \left(u_1 + u_2\right)x + \left(u_2 + u_1 - 6u_0\right)x^2 + ...$$

$$= 0 + 2x + 2x^2 + 3x^3 + 4x^4 + ... \qquad (2)$$

In the last step values have been substituted for $u_n + u_{n-1} - 6u_{n-2}$.
For example $u_3 + u_2 - 6u_1 = 3$.

The process now depends on your being able to sum the right-hand side of equation (2). The difficulty depends on the complexity of $f(n)$.

You should recognise (from Exercise 15E , Question 3) that

$$1 + 2x + 3x^2 + 4x^3 + ... = \frac{1}{(1-x)^2}.$$

So $\left(1+x-6x^2\right)G(x) = x\left(2+2x+3x^2+4x^3+...\right)$

$$= 2x + x\left(2x+3x^2+4x^3+...\right)$$

$$= 2x + x\left(\frac{1}{(1-x)^2} - 1\right)$$

$$= 2x + \frac{x}{(1-x)^2} - x$$

$$= x + \frac{x}{(1-x)^2}.$$

$\Rightarrow (1+3x)(1-2x)G(x) = \dfrac{x(1-x)^2 + x}{(1-x)^2}$

$$= \frac{x^3 - 2x^2 + 2x}{(1-x)^2}$$

$\Rightarrow \qquad\qquad G(x) = \dfrac{x^3 - 2x^2 + 2x}{(1+3x)(1-2x)(1-x)^2}.$

This result has to be reduced to partial fractions.

Let $G(x) \equiv \dfrac{A}{1+3x} + \dfrac{B}{1-2x} + \dfrac{C}{(1-x)^2} + \dfrac{D}{1-x}$

$$\equiv \dfrac{x^3 - 2x^2 + 2x}{(1-3x)(1-2x)(1-x)^2},$$

then solving in the usual way gives $A = -\frac{5}{16}$, $B = 1$, $C = -\frac{1}{4}$, $D = -\frac{7}{16}$.

Thus $G(x) = \dfrac{-5}{16(1+3x)} + \dfrac{1}{(1-2x)} - \dfrac{1}{4(1-x)^2} - \dfrac{7}{16(1-x)}$

$$= -\tfrac{5}{16}\left(1 - 3x + (-3x)^2 + \ldots\right) + \left(1 + 2x + (2x)^2 + \ldots\right)$$

$$-\tfrac{1}{4}\left(1 + 2x + 3x^2 + \ldots\right) - \tfrac{7}{16}\left(1 + x + x^2 + \ldots\right).$$

Picking out the coefficient of the nth term in each bracket gives

$$u_n = -\tfrac{5}{16}(-3)^n + 2^n - \tfrac{1}{4}(n+1) - \tfrac{7}{16}$$

$$= \tfrac{1}{16}\left[-5(-3)^n + 16 \times 2^n - 4(n+1) - 7\right].$$

As you can see, the form of u_n is still $Am_1^n + Bm_2^n$, but with the addition of a term of the form $Cn + D$. This additional term is particular to the function $f(n)$, which was n in this case, and to the values of u_0 and u_1.

Exercise 15F

1. Write the following as partial fractions.

   (a) $\dfrac{1}{(1-x)(1-2x)}$

   (b) $\dfrac{2x-3}{(2-x)(1+x)}$

   (c) $\dfrac{x^2+2}{(x+1)(1-2x)^2}$

2. Sum these series. Each is the result of expanding the expression of the form $(a+bx)^n$ using the Binomial Theorem (commonly $n = -1$ or -2).

   (a) $1 + x + x^2 + x^3 + \ldots$

   (b) $1 + 2x + 3x^2 + 4x^3 + 5x^4 + \ldots$

   (c) $-2 - 3x - 4x^2 - 5x^3 - 6x^4 - \ldots$

   (d) $x^2 + 2x^3 + 3x^4 + 4x^5 + \ldots$

   (e) $1 + 2x + (2x)^2 + (2x)^3 + (2x)^4 + \ldots$

   (f) $(5x)^2 + (5x)^3 + (5x)^4 + \ldots$

3. Expand as power series :

   (a) $(1-3x)^{-1}$ (b) $\dfrac{1}{(2-x)^2}$ (c) $\dfrac{x}{x-1}$.

   Give the nth term in each case.

4. Using the generating function method, solve

   (a) $u_n - 2u_{n-1} - 8u_{n-2} = 8$, $n \geq 2$, given $u_0 = 0$, $u_1 = 2$.

   (b) $u_n - 2u_{n-1} = 3^n$, $n \geq 1$, given $u_0 = 1$.

   (c) $u_n - u_{n-1} - 2u_{n-2} = n^2$, $n \geq 2$, given $u_0 = 0$, $u_1 = 1$.

15.7 Miscellaneous Exercises

1. Solve $u_n - 4u_{n-2} = 0$, $n \geq 3$, when $u_1 = 2$ and $u_2 = 20$.

2. Find the general solution of $u_n - 4u_{n-1} + 4u_{n-2} = 0$.

3. The life of a bee is quite amazing. There are basically three types of bee :

    queen   a fertile female

    worker an infertile female

    drone   a fertile male.

    Eggs are either fertilised, resulting in queens and workers or unfertilised, resulting in drones.

    Trace back the ancestors of a drone. Find the numbers of ancestors back to the nth generation. The generation tree has been started for you below :

    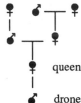

    queen

    drone

4. Find the nth term of these sequences :

    (a) 2, 5, 11, 23, ...

    (b) 2, 5, 12, 27, 58, ...

    (c) 1, 2, 6, 16, 44, ...

5. Solve the difference equation

    $$u_n - u_{n-1} - 12u_{n-2} = 2^n, \quad n \geq 2,$$

    if $u_0 = 0$, and $u_1 = 1$.

6. Write down the general solution of the model of the economy in Activity 4 when $A = \frac{2}{3}$, $B = 4$ and $N_1 = 1$, $N_2 = 2$.

7. A difference equation of the form

    $$u_n + au_{n-1} + bu_{n-2} = k$$

    defines a sequence with its first five terms as 0, 2, 5, 5, 14. Find the nth term.

8. Find the smallest value of n for which u_n exceeds one million if $u_n = 10 + 3u_{n-1}$, $n \geq 1$, given $u_0 = 0$.

9. Find the solution of the difference equation

    $$u_n = u_{n-2} + n, \quad n \geq 2,$$

    given $u_1 = u_0 = 1$.

10. (a) Solve, by iteration, the recurrence relation

    $$u_n = 4 - 3u_{n-1},$$

    (i)   subject to the initial condition $u_1 = 10$;

    (ii) subject to the initial condition $u_1 = 1$.

    (b) A ternary sequence is a sequence of numbers, each of which is 0, 1 or 2. (For example, 1002 and 1111 are 4-digit ternary sequences.) Let u_n be the number of n-digit ternary sequences which do not contain two consecutive 0s. By considering the number of such sequences which begin with 0, and the number which begin with 1 or 2, find a second-order linear recurrence relation for u_n, and write down appropriate initial conditions.

    (c) Using the method of generating functions, solve the recurrence relation

    $u_n - 5u_{n-1} + 6u_{n-2} = 0$, subject to the initial conditions $u_0 = 1$, $u_1 = 3$.

11. Write down the general solution of the difference equation

    $$u_{n+1} = 3u_n, \quad n \geq 1.$$

    Hence solve the difference equation

    $$u_{n+1} = 3u_n + 5, \text{ given that } u_1 = 6.$$

12. In a new colony of geese there are 10 pairs of birds, none of which produce eggs in their first year. In each subsequent year, pairs of birds which are in their second or later year have, on average, 4 eggs (2 male and 2 female)). Assuming no deaths, show that the recurrence relation which describes the geese population is

    $$u_{r+1} = u_r + 2u_{r-1}, \; u_1 = 10 \text{ and } u_2 = 10,$$

    where u_r represents the geese population (in pairs) at the beginning of the rth year.

13. A pair of hares requires a maturation period of one month before they can produce offspring. Each pair of mature hares present at the end of one month produces two new pairs by the end of the next month. If u_n denotes the number of pairs alive at the end of the nth month and no hares die, show that u_n satisfies the recurrence relation $u_n = u_{n-1} + 2u_{n-2}$.

    Solve this recurrence relation subject to the initial conditions $u_0 = 6$ and $u_1 = 9$.

    Also find the solution for u_n if the initial conditions $u_0 = 4$ and $u_1 = 8$. In this case, how many months will it take for the hare population to be greater than 1000?

14. Find the general solution of the recurrence relation $u_n + 3u_{n-1} - 4u_{n-2} = 0$.

15. (a) Solve the recurrence relation $u_n = (n-1)u_{n-1}$, subject to the initial condition $u_1 = 3$.

    (b) Find the general solution of the recurrence relation $u_n + 4u_{n-1} + 4u_{n-2} = 0$.

16. In an experiment the pressure of gas in a container is measured each second and the pressure (in standard units) in n seconds is denoted by p_n. The measurements satisfy the recurrence relation $p_0 = 6$, $p_1 = 3$, and

$p_{n+2} = \dfrac{1}{2}(p_{n+1} + p_n)$ $(n \geq 0)$. Find an explicit

formula for p_n in terms of n, and state the value to which the pressure settles down in the long term.

17. The growth in number of neutrons in a nuclear reaction is modelled by the recurrence relation $u_{n+1} = 6u_n - 8u_{n-1}$, with initial values $u_1 = 2$, $u_2 = 5$, where u_n is the number at the beginning of the time interval n $(n = 1, 2, ...)$. Find the solution for u_n and hence, or otherwise, determine the value of n for which the number reaches 10 000.

(AEB)

18. A population subject to natural growth and harvesting is modelled by the recurrence relation $u_{n+1} = (1+\alpha)u_n - k2^n$. Here u_n denotes the population size at time n and α and k are positive numbers. If $u_0 = a$, find the solution for u_n in terms of n, α, a and k.

(a) With no harvesting $(k=0)$, $a=100$ and $\alpha = 0.2$, determine the smallest value of n for which $u_n \geq 200$.

(b) With $k=1$, $a=200$ and $\alpha=0.2$,

  (i) Show that $u_n = (201.25)(1.2)^n - (1.25)2^n$

  $(n = 0, 1, 2, ...)$

  (ii) What is the long term future of the population?

  (iii) Determine the value of n which gives the greatest population.

(AEB)

APPENDIX 1

Braille patterns

Letters and numbers

1	2	3	4	5	6	7	8	9	0
a	b	c	d	e	f	g	h	i	j

k	l	m	n	o	p	q	r	s	t

u	v	w	x	y	z

Punctuation

comma (,)	apostrophe (')	semicolon (;)	colon (:)	period (.)

hyphen (-)	letter sign	capital sign	exclamation (!)	number sign

decimal point	question mark ?	parenthesis ( )	oblique stroke /	accent sign

APPENDIX 2

Left hand codes

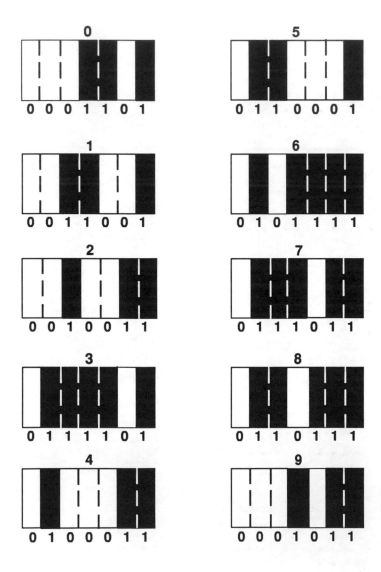

0 — 0 0 0 1 1 0 1	**5** — 0 1 1 0 0 0 1
1 — 0 0 1 1 0 0 1	**6** — 0 1 0 1 1 1 1
2 — 0 0 1 0 0 1 1	**7** — 0 1 1 1 0 1 1
3 — 0 1 1 1 1 0 1	**8** — 0 1 1 0 1 1 1
4 — 0 1 0 0 0 1 1	**9** — 0 0 0 1 0 1 1

APPENDIX 3

Number sets

Value of digit	Number set A (odd)	Number set B (even)	Number set C (even)
0			
1			
2			
3			
4			
5			
6			
7			
8			
9			

APPENDIX 4

Coding system for left most digit of EAN 13

Value of digits	Number sets for coding left hand numbers
0	A A A A A A
1	A A B A B B
2	A A B B A B
3	A A B B B A
4	A B A A B B
5	A B B A A B
6	A B B B A A
7	A B A B A B
8	A B A B B A
9	A B B A B A

APPENDIX 5

Codes

Code 1

0 0 0 0
0 0 1 1
0 1 0 1
0 1 1 0
1 0 0 1
1 0 1 0
1 1 0 0
1 1 1 1

Code 2

0 0 0 0 0 0
0 0 1 1 1 0
0 1 0 1 0 1
0 1 1 0 1 1
1 0 0 0 1 1
1 0 1 1 0 1
1 1 0 1 1 0
1 1 1 0 0 0

Code 3

0 0 0 0
0 1 0 1
1 0 1 0
1 1 1 1

Code 4

0 0 0 0
1 1 0 0
0 0 1 1
1 1 1 1

Code 5

0 0 0 0 0 0 0
0 0 1 1 1 0 1
0 1 0 1 0 1 1
0 1 1 0 1 1 0
1 0 0 0 1 1 1
1 0 1 1 0 1 0
1 1 0 1 1 0 0
1 1 1 0 0 0 1

Code 6

0 0 0 0 0 0 0
1 1 0 1 0 0 1
0 1 0 1 0 1 0
1 0 0 0 0 1 1
1 0 0 1 1 0 0
0 1 0 0 1 0 1
1 1 0 0 1 1 0
0 0 0 1 1 1 1
1 1 1 0 0 0 0
0 0 1 1 0 0 1
1 0 1 1 0 1 0
0 1 1 0 0 1 1
0 1 1 1 1 0 0
1 0 1 0 1 0 1
0 0 1 0 1 1 0
1 1 1 1 1 1 1

Code 7

0 0 0 0 0 0 0 0
0 0 0 0 1 1 1 1
0 0 1 1 0 0 1 1
0 1 0 1 0 1 0 1
0 1 1 0 0 1 1 0
0 1 0 1 1 0 1 0
0 0 1 1 1 1 0 0
0 1 1 0 1 0 0 1
1 1 1 1 1 1 1 1
1 1 1 1 0 0 0 0
1 1 0 0 1 1 0 0
1 0 1 0 1 0 1 0
1 0 0 1 1 0 0 1
1 0 1 0 0 1 0 1
1 1 0 0 0 0 1 1
1 0 0 1 0 1 1 0

Code 8

0 0 0 0 0 0 0
0 0 1 0 1 1 1
0 1 0 0 1 1 0
0 1 1 0 0 0 1
1 0 0 0 1 0 1
1 0 1 0 0 1 0
1 1 0 0 0 1 1
1 1 1 0 1 0 0
0 0 0 1 0 1 1
0 0 1 1 1 0 0
0 1 0 1 1 0 1
0 1 1 1 0 1 0
1 0 0 1 1 1 0
1 0 1 1 0 0 1
1 1 0 1 0 0 0
1 1 1 1 1 1 1

Code 9

0 1 0 1 0 1 0
1 0 0 1 1 0 0
0 0 1 1 0 0 1
1 1 1 0 0 0 0
0 1 0 0 1 0 1
1 0 0 0 0 1 1
0 0 1 0 1 1 0

ANSWERS

The answers to the questions set in the Exercises are given below. Answers to questions set in some of the Activities are also given where appropriate.

1 GRAPHS

Exercise 1A

1. (a) (i) 4 (ii) 4 (iii) 2, 2, 2, 2
   (b) (i) 4 (ii) 6 (iii) 2, 2, 4, 4
   (c) (i) 4 (ii) 5 (iii) 2, 2, 2, 4
   (d) (i) 6 (ii) 15 (iii) 5, 5, 5, 5, 5, 5
   (e) (i) 2 (ii) 1 (iii) 1, 1
   (f) (i) 5 (ii) 4 (iii) 1, 1, 2, 2, 2

2. (i) (a), (d), (e) and (f)
   (ii) (a), (b), (c), (d) and (e)
   (iii) (d) and (e)

3. There are many possible solutions.

Exercise 1B

(a) and (f), (b) and (d), (c) and (k), (g) and (i), (j) and (l) are isomorphic.

Activity 1

Vertices	1	2	3	4	5	6	7	8
Graphs	1	2	4	11	34	156	1044	12346

Exercise 1C

1. (a) PAQ, PBQ, PCQ, PABQ, PBAQ, PBCQ, PCBQ, PABCQ, and PCBAQ.
   (b) PABPCQ, PABCPBQ, PAQCBQ, ...
   (c) PABPAQ, PABPCBAQ, PAQBAPCQ, ...
   (d) PABP, PBCP, PABCP, PAQBP, PAQCP, PBQCP, PABQCP, PBAQCP, and their reverses.

2. (a) only

Exercise 1D

(b) and (e) are Eulerian; (d) and (f) are semi-Eulerian.

Exercise 1E

(a), (b), (d) Hamiltonian

(c), (e), (f) not Hamiltonian

Activity 7

Vertices	1	2	3	4	5	6	7	8	9	10
Trees	1	1	1	2	3	6	11	23	47	106

Exercise 1F

1. 3
2. 6

Miscellaneous Exercises

2. 5

3. ST, SPT, SQT, SRT, SPQT, SPRT, SQPT, SQRT, SRPT, SRQT, SPQRT, SPRQT, SQRPT, SQPRT, SRQPT and SRQPT.

4. (c) is Eulerian and (a) is semi-Eulerian.

6. 9

11. (a) G_2 is Eulerian (b) Neither is Eulerian

12. (a) 0, 2 or 4 (b) 2 or 4 (c) 2 or 4

13. (a) No (b) Yes, AEBCFDA etc.

2 TRAVEL PROBLEMS

Exercise 2A

1. S-K-L-T = 54; S-L-N-T = 29; S-P-X-T = 16
2. A-Q-R-B = £15
3. M-C-B-D-N = 170 minutes. (Add 10 minutes to each of the journey times not starting at M.)

Exercise 2B

1. 155 m
2. 180 m
3. 500 ECU

Exercise 2C

1. 63, 36
2. 1700 m
3. 1050 m

Exercise 2D

1. A-B-C-E-D-A = 30
2. E-S-A-F-I-T-A-P-G-E = 917 miles, but the non-Hamiltonian route, E-G-A-S-A-F-I-T-A-P-E, gives just 861 miles
3. O-B-A-D-C-O = 29 minutes

Exercise 2E

1. 28; 92; 71
2. 740 m
3. 34 km

Miscellaneous Exercises

1. 35
2. 21
3. 108
4. 46
5. 91 km
6. BR-QL-SW-QX-BA-SO-TA-BA-BR; 67 miles
7. 316 miles
8. $250
9. 543 feet
10. Yes - 26 hours travelling plus 48 hours visiting.
11. 17
12. 39
13. (a) 271
    (b) 1015
14. 78
15. (a) 341 km   (b) ABFEDCA = 475 km
    (c) ABCDEFA = 583 km
16. (a) ACDE are odd.
    (b) S-C-D-C-E-D-F-E-A-E-S-G-A-S;
        AE and CD are repeated; 584 km
    (c) No. SG, GA and AS could be traversed in opposite directions.
17. (a) Lines C, D, D, A, A; 24 mins
    (b) Line A without changing; 27 mins
18. (a) M-B-L-P-C; 220 min
    (b) M-L-C

3 ENUMERATION

Exercise 3A

1. 12
2. 700 000
3. 9
4. 162
5. 46 620

Exercise 3B

1. 40 320
2. 6.40×10^{15}

3. 6.23×10^9
4. 420
5. 37 837 800
6. 10 080
7. 60, 36
8. 336
9. 12
10. 1 058 400

Activity 4

(a) The point of the Round Table, according to the legend, was that all the places were equal. Where-ever Arthur sits automatically becomes the head of the table, and all that remains is for the twelve knights to seat themselves in 12! ways.

(b) If there is a clasp, ten beads can be threaded in 10! ways. However, the same necklace can be turned over to give what appears to be a different order, so in reality there are only 10!÷2 different necklaces. If there is no clasp, the beads can slide freely around the loop of thread. In that event, the problem is analogous to the round table problem (see (a) above) with the extra point that the necklace can be turned over. On these assumptions there are 9!÷2 possibilities.

Exercise 3C

1. (a) 120   (b) 210   (c) 3060
   (d) 1   (e) 1
2. 1540
3. 56
4. 2.61×10^8
5. 5.22×10^{10}
6. 43 263
7. 4004
8. 6435
9. 10 weeks
10. 19 757 815

Exercise 3D

1. 30
2. 2454
3. 44
4. 477
5. 960

Exercise 3E

1. $\dfrac{7}{102}$

2. $\dfrac{2}{11}$

3. $\dfrac{7}{15}$

4. $\dfrac{11}{4165}$

5. 4.47×10^{-28}

Exercise 3F

1. 63
2. 33 554 431
3. 255
4. 30
5. 13699 – here the order of the letters is important and the 'quick method' does not work.

Exercise 3G

1. Divide the square into 25 smaller squares each with side 1.4 cm.
2. Each number (apart from 1) has a least prime factor from $\{2, 3, 5, 7, 11, 13, 17, 19\}$.
3. The midpoint is a lattice point if x_1 and x_2 are both even or both odd, and y_1 and y_2 are both even or both odd.
4. Build up the set one member at a time; at each stage, look at the remainder when the total is divided by n. Either one of these remainders is 0, or there are two the same.
5. Point A is joined to 16 other points by lines of six colours; there must be at least six points all joined to A by the same colour. Then consider how these six points are joined to one another, as in the second example.

Exercise 3H

1. 8 347 680
2. 9
3. 200
4. 207 774
5. 629 070

Exercise 3I

1. 286
2. 36
3. 171
4. 43 758
5. 55

Activity 8

(a) 28
(b) 891

Activity 9

Number	2	3	4	5	6	7	8	9	10
Partitions	1	2	4	6	10	14	21	29	41

Activity 10

Number of objects	1	2	3	4	5
Derangemants	0	1	2	9	44

Activity 11

Number of objects	6	7	8	9	10
Derangements	265	1854	14833	133496	1334961

Activity 12

As $n \to \infty, \dfrac{d_n}{n!} \to \dfrac{1}{e} \approx 0.3679$. The proof is beyond the scope of this course.

Miscellaneous Exercises

1. 40 320
2. 220
3. 126
4. 4 989 600
5. 13 860
6. 480
7. 12 144
8. 105
9. 54
10. 0.2637...
11. 40 320
12. 176
13. 458
14. 204 squares, 1296 rectangles
15. There are only ten possible units digits, so if there are eleven numbers there must be at least two with the same units digit. Their difference is divisible by 10.
16. If the 'round table' assumption is applied, suppose Mrs A's seat is the head of the table. The other three wives can sit in the three 'alternate' seats, in $3!=6$ ways. Mr A now has a choice of two seats (because there are two next to Mrs A) and a diagram will show that the other three men then have no choice at all. So there are just 12 ways of seating the couples according to these rules.
17. This is a partition problem; five dice have to be divided between six scores, and this can be done in $\binom{10}{5} = 252$ ways.
18. There are 49 possibilities. The result can be found by listing, but there is an alternative approach. By common sense, with just 50p and 20p coins the number of ways of making up various totals is as shown in the table.

Total/p	0	10	20	30	40	50	60	70	80	90	100
Ways	1	0	1	0	1	1	1	1	1	1	2

If 10p coins are also allowed, the number of ways of making a total Xp is the number of ways of making any total *less than or equal to* Xp using only 50p and 20p coins, because the difference can then be made up with 10p coins. This gives the new table

Total/p	0	10	20	30	40	50	60	70	80	90	100
Ways	1	1	2	2	3	4	5	6	7	8	10

With 5p coins as well, the same principle applies; the number of ways of making Xp is the number of ways of making any total less than or equal to Xp with just 50p, 20p and 10p coins, because 5p coins can be used to make up the balance. Hence the final table.

Total/p	0	10	20	30	40	50	60	70	80	90	100
Ways	1	2	4	6	9	13	18	24	31	39	49

19. If there is no restriction at all, 30 marks can be distributed between four questions in $^{33}C_3 = 5456$ ways, but some of these distributions give more marks to one question than the maximum allowed. If Q.1 is given at least 21 marks, the remaining 9 marks can be distributed between the four questions (including Q.1) in $^{12}C_3 = 220$ ways. If Q.2 is given at least 11 marks there are another $^{22}C_3 = 1540$ invalid ways, and similarly 1540 for Q.3 and 1540 for Q.4. But if Q.2 and Q.3 are both invalid, the remaining 8 marks can be shared in $^{11}C_3 = 165$ ways, and similarly for Q.2/ Q.4 and Q.3/Q.4. So the valid ways of scoring 30 marks are

$$5456 - (220 + 1540 + 1540 + 1540) + (165 + 165 + 165)$$
$$= 1111.$$

20. Consider a particular element X. If X is in a subset alone, then the other $(n-1)$ elements are in the other $(k-1)$ subsets, which is possible in $S(n-1, k-1)$ ways. On the other hand, if X is in a subset with other elements, the other $(n-1)$ elements are in k subsets, which is possible in $S(n-1, k)$ ways, and there are k choices of which subset X joins. Hence the proof. Clearly $S(n,1) = S(n,n) = 1$ for any value of n, and step-by-step application of the recurrence relation gives $S(6,3) = 90$.

21. (a) 3654   (b) 3654   (c) 5984

4   INEQUALITIES

Activity 2

If $x > y$ and $k < 0$, then $kx < ky$.

Exercise 4A

3. $x \le 2$

4. (a) $\{x \ge 2\}$ (b) $\{-2 \le x < -1\} \cup \{x > 1\}$

5. $\{x \le 1\} \cup \{x \ge 4\}$

Exercise 4B

1. (a) $\{x \le -1\} \cup \{x \ge \frac{2}{3}\}$ (b) $\{-2 \le x \le 4\}$

5. Yes

Activity 4

$A \ge G \ge H$

Exercise 4C

1. (a) A = 2.5   G = 2.21   H = 1.92
   (b) A = 2.5   G = 1.31   H = 0.36
   (c) A = 26.27   G = 2.78   H = 0.37
   (d) A = 251.25   G = 1.57   H = 0.004
   (e) A = 750.00   G = 1.41   H = 0.0027

3. 2

Exercise 4D

1. (a) 9 cm^2 (b) 8 cm^2 (c) 5 cm^2

2. (a) $\dfrac{\pi}{3\sqrt{3}} \approx 0.605$ (b) $\dfrac{\sqrt{3}\,\pi}{6} \approx 0.907$

   (c) $\dfrac{2\pi}{9} \approx 0.698$

3. $\dfrac{\pi k}{(1+k)^2}$;   (a) $k = 1$ (b) $k = 0$

4. $\dfrac{4\sqrt{3}}{\sqrt{\pi}} \approx 3.909 \text{ cm}^3$

5. $2\sqrt{2} \approx 2.828 \text{ cm}^3$

6. Sphere

Miscellaneous Exercises

1. (a) $\left\{-\frac{1}{\sqrt{2}} < x < 0\right\} \cup \left\{x > \frac{1}{\sqrt{2}}\right\}$

   (b) $\left\{-3 < x < -\sqrt{3}\right\} \cup \left\{-1 < x < \sqrt{3}\right\}$

2. $\frac{1}{2} < x < \frac{5}{2}$

3. $\{-1 < x < 1\} \cup \{x > 2\}$

4. $\left\{-\frac{3}{4} < x < 3\right\}$

5. For $x \le -1\frac{2}{3}$, $2x+5y<10$; $x \ge -1\frac{2}{3}$, $x+y<1$

6. No

8. Equality occurs when numbers are equal.

9. $\dfrac{4\pi\left(k+\dfrac{\pi}{8}\right)}{\left(1+2k+\dfrac{\pi}{2}\right)^2}$; $k=\frac{1}{2}$

10. For triangles, I.Q. $\le \dfrac{\pi}{3\sqrt{3}}$, equality only

occurring for equilateral triangles.

11. I.Q. ≤ 1, equality only occurs for a sphere.

5   LINEAR PROGRAMMING

Activity 1

Profit is £61; maximum profit of £66 at $x=6, y=3$.

Exercise 5A

1. If $x=$ no. of blouses, and $y=$ no. of skirts,

    maximise $P=8x+6y$
    subject to $x+y \le 7$

    $x+\frac{1}{2}y \le 5$

    $x \ge 0$ and $y \ge 0$.

2. If $x=$ no. of large vans, and $y=$ no. of small vans,

    minimise $P=40x+20y$
    subject to $5x+2y \ge 30$

    $2x+y \le 15$

    $x \le y$

    $x \ge 0$ and $y \ge 0$.

3. If $x=$ no. of boxes of wood screws, and $y=$ no. of boxes of metal screws,

    maximise $P=10x+17y$
    subject to $3x+2y \le 3600$

    $2x+8y \le 3600$

    $x \ge 0$ and $y \ge 0$.

4. If $x=$ no. of apprentices, and $y=$ no. of skilled workers,

    $x+2y < 180$

    $x+y \ge 110$

    $y \ge 40$

    $y \ge \frac{1}{2}x$

    $x \ge 0$.

Exercise 5B

1. $x=3$, $y=4$ and $P=£48$

2. $x=4$, $y=5$ and $P=£260$

3. $P=£18$
    (a) 6 caravans, 6 tents
    (b) 3 caravans, 12 tents

4. 40 adults and 10 junior members; 46 members

5. $X=\frac{12}{11}$ kg, $Y=\frac{40}{11}$ kg; any point on the line

    between $\left(\frac{12}{11}, \frac{40}{11}\right)$ and $(12, 0)$

Activity 4

B : $r=t=0$

C : $s=r=0$

D : $y=s=0$

Activity 5

$P = 1600 - \frac{10}{3}s + \frac{160}{3}y$

$P = 2240 + 2s - 160r$

$P = 2320 - 80r - 20t$

Exercise 5C

1. $P=\frac{188}{19}$ at $x=\frac{30}{19}$, $y=\frac{32}{19}$,

2. $P=\frac{43}{11}$ at $x=\frac{28}{11}$, $y=\frac{15}{11}$

3. $P=56$ at $x=2$, $y=\frac{12}{5}$

Exercise 5D

1. $P=21$ at $x=0$, $y=\frac{7}{2}$

2. $P=45$ at $x=0$, $y=\frac{5}{2}$, $z=\frac{15}{8}$

3. $P=8$ at $x=0$, $y=1$, $z=0$

4. $\frac{12}{5}$ at $x=\frac{2}{5}$, $y=\frac{1}{5}$, $z=0$

Miscellaneous Exercises

1. Profit = 13860 at $x=1080$, $y=180$.

2. Type A = 818, Type B = 163 with profit of £376.10 (best integer solution)

3. 19; no. of skilled workers = 11; no. of apprentices = 10.

4. No. of small spaces = 48 and no. of large spaces = 24 giving a total 72.

5. 1270 tonnes when 12 Type A and 13 Type B are used.

6. 25 trees; 11 trees

7. Upper seam = $\frac{36}{7}$ h, and lower seam = $\frac{5}{7}$ h.

8. (a) $P = 20x + 30y$; $3x + 5y \leq 225, 2x + y \leq 80$

   (b) £1400; $x = 25$, $y = 30$

   (c) £1350

9. (a) $6x + 2y \geq 18$, $3x + 3y \geq 21$, $2x + 4y \geq 16$

   (c) 1 Xtravit and 6 Yeastalife

10. $S = 20$, $T = 20$, $C = 0$; $P = 2800$

11. $\dfrac{18}{11}$, $\dfrac{40}{11}$; cost $12\dfrac{2}{11}$p

6 PLANAR GRAPHS

Exercise 6A

1. K_4 is planar; K_6 is non-planar. If $n \geq 5$ then K_n is non-planar.

2. n odd

3. Tetrahedron, cube and octahedron.

4. (a) 10   (b) $\frac{1}{2}n(n-1)$

5. Yes

Exercise 6B

2. rs

4. r and s both even; one of r, s is odd and the other is 2

Exercise 6C

1. (a) Possible (b) Possible

Exercise 6D

1. G_1

Miscellaneous Exercises

2. $r > 2$ and $s > 2$

3. (a) 1   (b) 3   (c) 0

6. Non-planar

7 NETWORK FLOWS

Activity 1

600 cars per hour.

Extra capacity from D to A would improve flow.

Exercise 7A

1. Arc AB is over capacity; inflow and outflow are not equal for vertex C.

Exercise 7B

1. 16 (A - E)

2. (a) 26 (b) 16 (c) 53

Exercise 7C

1. 18, 57, 38

2. 75

Exercise 7D

1. (a) 28 (b) 23 (c) 111

2. 16

Exercise 7E

1. 19

2. No; No

3. 18

Miscellaneous Exercises

1. $N_1 - 31$; $N_2 - 15$; $N_3 - 29$

2. 46

3. (a) No (b) Yes (c) No (d) Yes; 18 for (b) and (d)

4. 42

5. 5200

6. 1800

7. $8 \times 500 = 4000$

8. (b) 18

   (c) CF, CE, SB, DB, DT

9. (b) 22

10. (c) 9

11. (c) 7

13. (b) 135

14. (c) 9

15. (a) $k = 9$

    (b) Yes

16. (a) 750, BT = 300, CT = 350, DT = 100

    (b) 550

    (c) DT by at least 70

8 CODES IN EVERYDAY USE

Activity 1

(a) 6   (b) 15   (c) 20   (d) 15   (e) 6   (f) 1

Total 63

Activities 4 and 7

No, there are a number of possibilities.

Activity 8

20

Activity 10

About 400 million.

Activity 12

640 million

Activity 14

$2^7 = 128$

Exercise 8B

1. (a) MANATETEA   (b) MEETTEAMATTEN

2.
Letter	Tally	Frequency	Code
B	1	1	011110
C	11	2	01110
H	1111	4	01100
I	111	3	01101
M	1	1	011111
P	~~1111~~ ~~1111~~	10	00
S	~~1111~~ ~~1111~~ 1	11	1
U	~~1111~~ 111	8	010

A possible solution is shown below.

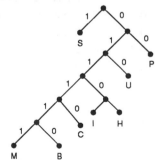

Miscellaneous Exercises

1. (a) One error in M gives N
   (b) - • - • - - is NKM or KJ

9 THEORY OF CODES

Activity 2

0 0 0 0 0, 1 1 1 0 0, 1 0 0 1 1, 0 1 1 1 1

Activity 3

(a) 0 0 0 0 0 0 (b) 1 1 1 0 0 0 (c) 1 1 0 0 1 1

Activity 4

Detect two errors; correct one error

Activity 5

$\delta = 2,\ 3,\ 2,\ 2,\ 4$

Exercise 9A

1. 10 codewords, $\delta = 2$

2. 35 codewords, $\delta = 2$;
   Detects 1 error but does not correct it.

3. $\delta = 4$
   Detects 2 errors; corrects 1 error.

Activity 9

1 1 1 1, 1 1 0 0, 0 0 1 1, 0 0 0 0

Activity 10

$$\begin{bmatrix} 1 & 0 & 1 & 0 \\ 1 & 1 & 1 & 1 \end{bmatrix} \text{ and } \begin{bmatrix} 0 & 1 & 0 & 1 \\ 1 & 1 & 1 & 1 \end{bmatrix}$$

Activity 11

$$\begin{bmatrix} 0 & 0 & 1 & 1 & 1 & 0 & 1 \\ 0 & 1 & 0 & 1 & 0 & 1 & 1 \\ 1 & 0 & 0 & 0 & 1 & 1 & 1 \\ 1 & 1 & 1 & 0 & 0 & 0 & 1 \end{bmatrix}$$

This is one possible answer; there are many others.

Activity 13

(a) 1 1 1 0 0 0 1 (b) 0 1 0 1 0 1 1

Exercise 9B

1. (a) 4 (b) Detects 2 errors, corrects 1 error.
   (c) 1 1 0 1 1 0 0

2. 1 1 0 0, 0 0 0 0; 1

3. (a) 4   (b) (i) 1 (ii) 2   (c) 1 1 0 0 1 0 1

4. $$\begin{bmatrix} 0 & 0 & 0 & 1 & 1 & 1 \\ 0 & 1 & 1 & 0 & 0 & 1 \\ 1 & 0 & 1 & 0 & 1 & 1 \end{bmatrix}; \ 0 1 1 1 1 0$$

Miscellaneous Exercises

1. (a) 7, 4, $\dfrac{4}{7}$

   (b) 0 1 1 1 1 0 0

2. (a) 2
   (b) 1 1 1 1 1 1 1 1 1 1
   (c) Yes

3. (a) 0100111, 1010011, 1101001, 1110100,
        0111010, 0011101

    (b) Hamming distance = 4

    (c) One error can be detected and corrected
         since that will leave a string distance 1
         from correct word and at least 3 from all
         others.

    (d) Not linear since any word + itself gives
         the zero word, not in code. Adding the
         zero word alone makes it linear.

    (e) Break message up into strings of 7,
         multiply matrix by these 7-strings as
         columns to give 110, 000, 011 respectively.

         Deduce that errors are in 3rd place,
         nowhere, 6th respectively to give correct
         version 010011111101000011101.

4. (a) 5, 2
    (b) 0010010111

10 LOGIC

Activity 2

PM	MP	PM
SM	MS	MS
S P	S P	S P

Activity 3

(a) I (b) V (c) I (d) V

Exercise 10A

1. (a) $p \wedge q$ (b) $\sim p \wedge \sim q$ (c) $p \oplus q$

    (d) $\sim(p \wedge q)$ or $\sim p \vee \sim q$

2. (a) The cooker is working but the visitors are
         not hungry.

    (b) There is enough food and the visitors are
         hungry but the cooker is not working.

    (c) Either the visitors are hungry or there is
         not enough food.

    (d) Either the cooker is working and there is
         enough food or the visitors are not hungry.

    (e) There is not enough food and either the
         cooker is not working or the visitors are
         not hungry.

Exercise 10B

1.
q	r	q∨r
0	0	0
0	1	1
1	0	1
1	1	1

2.
p	r	~p∧r
0	0	0
0	1	1
1	0	0
1	1	0

3.
p	r	p∨~r
0	0	1
0	1	0
1	0	1
1	1	1

4.
p	q	~p∨~q
0	0	1
0	1	1
1	0	1
1	1	0

Exercise 10C

1.
a	b	c	(a∨b)∨c
0	0	0	0
0	0	1	1
0	1	0	1
0	1	1	1
1	0	0	1
1	0	1	1
1	1	0	1
1	1	1	1

2.
a	b	c	a∧(b∧c)
0	0	0	0
0	0	1	0
0	1	0	0
0	1	1	0
1	0	0	0
1	0	1	0
1	1	0	0
1	1	1	1

3.
a	b	c	a∨(b∨c)
0	0	0	0
0	0	1	1
0	1	0	1
0	1	1	1
1	0	0	1
1	0	1	1
1	1	0	1
1	1	1	1

4.
a	b	c	(a∧b)∧c
0	0	0	0
0	0	1	0
0	1	0	0
0	1	1	0
1	0	0	0
1	0	1	0
1	1	0	0
1	1	1	1

5.
a	b	c	a∧(b∨c)
0	0	0	0
0	0	1	0
0	1	0	0
0	1	1	0
1	0	0	0
1	0	1	1
1	1	0	1
1	1	1	1

6.
a	b	c	(a∧b)∨(a∧c)
0	0	0	0
0	0	1	0
0	1	0	0
0	1	1	0
1	0	0	0
1	0	1	1
1	1	0	1
1	1	1	1

7.
a	b	c	a∨(b∧c)
0	0	0	0
0	0	1	0
0	1	0	0
0	1	1	1
1	0	0	1
1	0	1	1
1	1	0	1
1	1	1	1

8.
a	b	c	(a∨b)∧(a∧c)
0	0	0	0
0	0	1	0
0	1	0	0
0	1	1	1
1	0	0	1
1	0	1	1
1	1	0	1
1	1	1	1

Activity 6

a	b	a ⇒ b
0	0	1
0	1	1
1	0	0
1	1	1

Exercise 10D

1. (a) 1 (b) 1 (c) 0 (d) 0 (e) 1

2. (a) $b \Rightarrow a$ (b) $\sim c \wedge \sim b \wedge \sim a$ (c) $c \Rightarrow a$

   (d) $b \wedge \sim c \Rightarrow a$ (e) $\sim b \Rightarrow c \wedge a$

3. (a) Either I water the plants and the crops grow or I spread manure and the crops grow.

   (b) If I either spread manure or do not water the plants, then the crops do not grow.

   (c) If the crops grow then I water the plants and spread manure.

   (d) If the crops do not grow or I spread the manure then I water the plants.

Exercise 10E

1. (a) T (b) F

2. (a) If and only if the theme park has excellent rides and the entrance charges are not high, then attendances are large.

   (b) If the attendances are not large or the entrance charges are not high, then the theme park has excellent rides.

Exercise 10F

1. Contradiction
2. Contradiction
3. Tautology

Exercise 10G

1. Yes
2. No
3. Yes
4. No

Miscellaneous Exercises

1. (a) $a \wedge b$ (b) $a \wedge b$ (c) $a \wedge b$

   (d) $\sim a \wedge \sim b$ (e) $a \Rightarrow b$ (f) $\sim b \Rightarrow \sim a$

   (g) $a \wedge \sim b$ (h) $\sim b \Rightarrow a$

2. (a)

p	q	$(p \vee \sim q) \Rightarrow q$
0	0	0
0	1	1
1	0	0
1	1	1

(b)

p	q	$[p \vee (\sim p \vee q)] \vee (\sim p \wedge \sim q)$
0	0	1
0	1	1
1	0	1
1	1	1

(c)

p	q	$(\sim p \vee \sim q) \Rightarrow (p \wedge \sim q)$
0	0	0
0	1	0
1	0	1
1	1	1

(d)

p	q	$\sim p \Leftrightarrow q$
0	0	0
1	0	1
0	1	1
1	1	0

(e)

p	q	r	$(\sim p \wedge q) \vee (r \wedge p)$
0	0	0	0
0	0	1	0
0	1	0	1
0	1	1	1
1	0	0	0
1	0	1	1
1	1	0	0
1	1	1	1

(f)

p	q	$(p \Leftrightarrow q) \Rightarrow (\sim p \wedge q)$
0	0	0
0	1	1
1	0	1
1	1	0

3. (a) Yes (b) No (c) Yes (d) Yes
4. (a) No (b) Yes (c) No (d) No
5. (a) Yes (b) No (c) Yes (d) No
6. (a) Yes (b) No (c) Yes

11 BOOLEAN ALGEBRA

Exercise 11A

1.

x	y	$(\sim y \vee x)$	$x \wedge (\sim y \vee x)$
0	0	1	0
0	1	0	0
1	0	1	1
1	1	1	1

2.

a	b	c	$(\sim b \wedge c)$	$a \vee (\sim b \wedge c)$
0	0	0	0	0
0	0	1	1	1
0	1	0	0	0
0	1	1	0	0
1	0	0	0	1
1	0	1	1	1
1	1	0	0	1
1	1	1	0	1

3.

a	b	c	$(\sim b \vee c)$	$a \vee (\sim b \vee c)$	$[a \vee (\sim b \vee c)] \wedge \sim b$
0	0	0	1	1	1
0	0	1	1	1	1
0	1	0	0	0	0
0	1	1	1	1	0
1	0	0	1	1	1
1	0	1	1	1	1
1	1	0	0	1	0
1	1	1	1	1	0

4. $(a \wedge b) \vee \sim c$

5. $[\sim (p \wedge q) \vee (p \wedge r)] \wedge \sim r$

Exercise 11B

1. Equivalent
2. Equivalent
3. Equivalent
4. Not equivalent

Exercise 11C

1. $[A \wedge (B \vee D)] \vee (\sim D \wedge C)$

2. $A \wedge [(A \vee D \vee C) \vee (C \wedge (B \vee \sim D))]$

Exercise 11D

1. $a \vee b$
2. $a \wedge b$
3. $(a \wedge b) \vee (c \wedge d)$

Exercise 11E

1. $f(a,b) = (\sim a \wedge \sim b) \vee (a \wedge \sim b)$

2. $f(a,b) = (\sim a \wedge \sim b) \vee (\sim a \wedge b) \vee (a \wedge b)$

3. $f(x,y,z) = (\sim x \wedge \sim y \wedge \sim z) \vee (x \wedge \sim y \wedge \sim z) \vee (x \wedge y \wedge z)$

Exercise 11F

1. $a \wedge b = (a \uparrow b) \uparrow (a \uparrow b)$

2. $a \wedge \sim b = [a \uparrow (b \uparrow b)] \uparrow [a \uparrow (b \uparrow b)]$

3. $(\sim a \wedge \sim b) \vee \sim b = [(a \uparrow b) \uparrow (a \uparrow b)] \uparrow [(b \uparrow b) \uparrow (b \uparrow b)]$

Activity 4

a	b	c	Carry bit	Answer bit
0	0	0	0	0
0	0	1	0	1
0	1	0	0	1
0	1	1	1	0
1	0	0	0	1
1	0	1	1	0
1	1	0	1	0
1	1	1	1	1

Miscellaneous Exercises

1. (a)

a	b	c	Output
0	0	0	1
0	0	1	0
0	1	0	1
0	1	1	0
1	0	0	1
1	0	1	0
1	1	0	0
1	1	1	0

(b)

a	b	c	Output
0	0	0	1
0	0	1	0
0	1	0	1
0	1	1	0
1	0	0	1
1	0	1	0
1	1	0	1
1	1	1	1

(c)

a	b	c	Output
0	0	0	1
0	0	1	0
0	1	0	1
0	1	1	0
1	0	0	1
1	0	1	0
1	1	0	1
1	1	1	0

2. (a) $(\sim p \vee q) \wedge \sim r$

(b) $(\sim q \wedge r) \vee p$

(c) $[(\sim q \vee r) \wedge p] \vee q$

3. (a) $(A \vee B \vee C \vee \sim B) \wedge C$

(b) $[(P \vee Q) \wedge (R \vee P \vee S)] \vee (\sim Q \wedge \sim P)$

(c) $(A \wedge D) \vee (B \wedge E) \vee (A \wedge C \wedge E) \vee (B \wedge C \wedge D)$

4. (a) 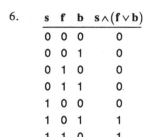   (b)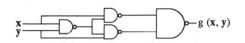

(b) $x \Rightarrow y$ equivalent to $\text{NAND}(\text{NAND}(x,y),x)$

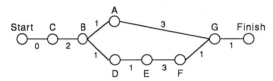

$$\sim(x \Leftrightarrow y) = \text{NAND}(x \Rightarrow y),(y \Rightarrow x)$$

6.

s	f	b	$s \wedge (f \vee b)$
0	0	0	0
0	0	1	0
0	1	0	0
0	1	1	0
1	0	0	0
1	0	1	1
1	1	0	1
1	1	1	1

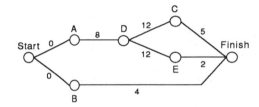

12 CRITICAL PATH ANALYSIS

Activity 1

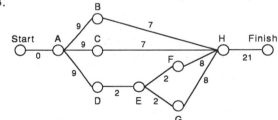

8. (a) $b \wedge c$ (b) $a \vee b$

(c) $(p \wedge q) \vee [(\sim p \vee \sim q) \wedge (r \vee s)]$

9. $[b \wedge (a \vee c)] \vee (a \wedge c)$

10. $[(a \uparrow a) \uparrow (a \uparrow a)] \uparrow (b \uparrow b)$

12. (a)

$x_1 x_2 x_3$	$\sim x_1$	$\sim x_1 \vee x_1$	$\sim x_3$	$\sim x_3 \vee x_2$	$\sim x_1 \vee x_2 \wedge \sim x_3 \vee x_2$
000	1	1	1	1	1
001	1	1	0	0	0
010	1	1	1	1	1
011	1	1	0	1	1
100	0	0	1	1	0
101	0	0	0	0	0
110	0	1	1	1	1
111	0	1	0	1	1

(b) $(\sim x_1 \vee x_2) \wedge (\sim x_3 \vee x_2) \equiv (\sim x_1 \wedge \sim x_3) \vee x_2$

$$\equiv \sim (x_1 \vee x_3) \vee x_2$$

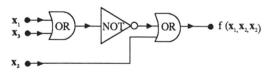

Exercise 12A

1. See Activity 1 above.

2.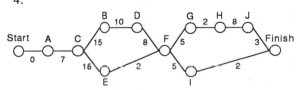

3.

4.

Activity 2

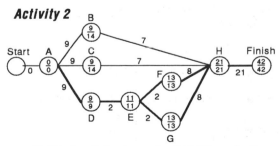

Critical path is shown by **bold** lines in the network.

13. (a)

p	q	$\sim (p \wedge q)$	$\sim (p \wedge q) \wedge p$	$\sim (\sim (p \wedge q) \wedge p)$	$p \Rightarrow q$
0	0	1	0	1	1
0	1	1	0	1	1
1	0	1	1	0	0
1	1	0	0	1	1

Last two columns the same, hence expressions equivalent.

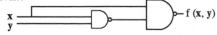

Exercise 12B

1. (a) Start - A - B - C - F - I - Finish
   (b) Start - A - B - D - E - G - Finish
   (c) Start - A - C - E - G - Finish
2. Start - A - C - B - D - F - G - H - J - Finish

Miscellaneous Exercises

1. (a) (i) 13 (ii) start - C - F - Finish
   (b) 2

2. (a)

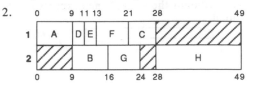

   (b) Start - B - C- D - G - H- Finish

   (c)

Activity	Latest Starting Time
A	8
B	0
C	4
D	12
E	22
F	15
G	18
H	28
I	16
J	20

3. (b) Start - A - D - H - Finish

   (c)

Activity	Starting Times	
	Earliest	Latest
A	0	0
B	4	5
C	0	1
D	4	4
E	0	4
F	4	13
G	4	8
H	9	9

4. (b) and (c)

Activity	Starting Times		Float	
	Earliest	Latest		
A	9	15	6	
B	12	13	1	
C	0	0	0	←
D	18	19	1	
E	9	9	0	←
F	0	4	4	
G	21	22	1	
H	7	10	3	
I	12	12	0	←
J	0	2	2	
K	21	21	0	←

   Critical path: Start - C - E - I - K - Finish

5. (b) 75 minutes
   (c) Start - A - B - C - E - F - G - I - Finish
7. (a) 12 weeks, Start - C - E - H - Finish
   (b) D = 3, B = 2
8.
Activity	A	B	C	D	E	F	G	H	I	Finish
Earliest start time	0	2	4	2	5	6	5	6	9	10
Latest start time	0	2	5	4	6	8	5	6	9	10

   (b) Minimum completion time is 10 days
       Critical path A B G H I
   (c) D and F, 2 days
9. (a) Critical path is B C F G
       Minimum completion time is 85 mins

13 SCHEDULING

Activity 1

0	2		7	12	17	22		32	34
1	A	H	I	J	K		L		B
2	C	D	E		F		G		
0	2	4		14		24		34	

Activity 2

Activity	Rank
A	0
B	15
C	15
D	15
E	7
F	7
G	7
H	4
I	4
J	4
K	4
L	14

This method gives a finishing time of $t = 24$, which is **not** optimal.

Exercise 13A

1.

0		4	9	15	16
1	A	D	G		F
2	C	E	B	H	
0	3	7	11		16

Optimum solution is produced.

2.

0	9	11	13	21	28		49
1	A	D	E	F	C		/////
2	/////	B		G		H	
0	9		16	24	28		49

Similar solution for second method.

These are optimal solutions for 2 workers.

3. For the problem in Activity 1, this method gives a completion time of 25 using 4 workers. It is not optimal.

Exercise 13B

1. (a)

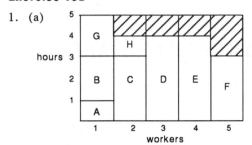

(b) 4, 4, 3, 3, 2, 2, 1, 1

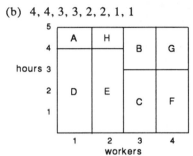

2.

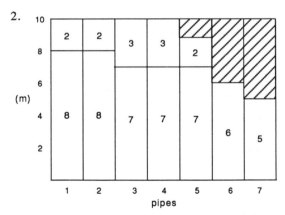

3. It is possible to meet the order with just 10 sheets, cut appropriately.

Exercise 13C

1. C and E, giving value 14.
2. A and C, giving income of £215 000.
3. C, D and E, giving value 18.

Miscellaneous Exercises

1. (a) 5 workers   (b) (i) 6   (ii) 5, yes
2. A, B, D, giving value 8.
3. Optimum solutions:
   (a) 4   (b) 4   (c) 5
4. (a) 5
   (b) 4

5. (a) 5
   (b) 4

14 DIFFERENCE EQUATIONS 1

Exercise 14A

1. (a) 11   (b) 58   (c) 12

2. (a) $u_n = u_{n-1} + 2$, $u_1 = 3$, $n \geq 2$

   (b) $u_n = 2u_{n-1} + 1$, $u_1 = 2$, $n \geq 2$

   (c) $u_n = 3u_{n-1} - 1$, $u_1 = 1$, $n \geq 2$

3. $u_n = \frac{1}{3} u_{n-1}$; between 12 and 13 strokes.

4. (a) 1, 4, 9, 16; $u_{n+1} = u_n + 2n + 1$, $n \geq 1$

   (b) 9, 17, 24, 30; $u_{n+1} = u_n + 9 - n$, $n \geq 1$

   (c) 9, 24, 45, 72; $u_{n+1} = u_n + 3(2n+3)$, $n \geq 1$

Exercise 14B

1. (a) $u_n = u_1 + 2(n-1)$

   (b) $u_n = 4^{n-1} u_1 + \frac{1}{3}\left(1 - 4^{n-1}\right)$

   (c) $u_n = 3^{n-1} u_1 + 3^{n-1} - 1$

2. $p_n = 1.025 p_{n-1}$; $p_n = (1.025)^n p_0$;
   $p_{20} \approx 819$ million;   16 - 17 years.

3. £24.96

Exercise 14C

1. (a) $u_n = 4^{n-1} u_1 + \frac{2}{3}\left(4^{n-1} - 1\right)$

   (b) $u_n = 4^n u_0 + \frac{2}{3}\left(4^n - 1\right)$

   (c) $u_n = 3^n u_0 - \frac{5}{2}\left(3^n - 1\right)$

   (d) $u_n = u_0 + 6n$

   (e) $u_n = u_1 - 8(n-1)$

   (f) $u_n = (-2)^n u_0 - \frac{4}{3}\left((-2)^n - 1\right)$

   (g) $u_n = (-3)^n u_0 - \frac{1}{2}\left((-3)^n - 1\right)$

   (h) $u_n = (-4)^n u_0 + \frac{3}{5}\left((-4)^n - 1\right)$

   (i) $u_n = 4^{n-1} u_1$

   (j) $u_n = 4^n u_0 - \frac{5}{3}\left(4^n - 1\right)$

2. (a) $u_n = \frac{1}{2}\left(7 \times 3^n - 5\right)$

   (b) $u_n = 2 + (-2)^{n-1}$

   (c) $u_n = 4 - 3n$

   (d) $u_n = \frac{3}{4}\left(5^{n-1} - 1\right)$

   (e) $u_n = 3 + 7n$

   (f) $u_n = \frac{3}{4}(-3)^n + \frac{1}{4}$

Exercise 14D

1. (a) $u_n = 3^{n-1}u_1 + 2\left(3^{n-1} - 1\right)$

   (b) $u_n = \left(\frac{1}{2}\right)^{n-1}u_1 - \left(\frac{1}{2}\right)^{n-3} + 4$

   (c) $q_n = 3n - 2$

   (d) $a_n = 2^{n+1}$

   (e) $b_n = 2 \times 4^{n-1} + \frac{5}{3}\left(4^{n-1} - 1\right)$

2. $u_n = 10 \times 2^n - 3$

3. £24.95

4. £75.35

6. 6165 tonnes; 15547 tonnes

7. Approximately 26 months

Exercise 14E

1. $u_n = \frac{1}{2}(n+2)(n-1) + 5$

2. (a) $u_n = u_1 + \frac{n}{6}(n+1)(2n+1) - 1$

   (b) $u_n = u_1 + 2^2\left(2^{n-1} - 1\right)$

   (c) $u_n = 2^{n-1}u_1 + 3 \times 2^{n-1} - (n+2)$

3. $k = 3$; $u_6 = 1577$

4. 126%

Miscellaneous Exercises

1. (a) $u_n = 2^{n-1}u_1$ (b) $u_n = 3^{n-1}u_1 + \frac{3}{2}\left(3^{n-1} - 1\right)$

   (c) $u_n = 3^{n-1}u_1 + \frac{5}{4} \times 3^{n-1} - \frac{1}{2}n - \frac{3}{4}$

2. $u_n = 3u_{n-1} - 2$; $u_n = 3^{n-1} + 1$; $u_{10} = 19684$

3. £60.28

4. 16.35 million

5. n; $u_n = u_{n-1} + n - 1$; $u_n = \frac{n(n-1)}{2}$; $u_{20} = 190$

6. $p = 4$, $q = -5$, $u_6 = 343$

7. (a) £315.96   (b) £340

8. 37 million

9. 47 months

10. Yes

12. $u_n = \dfrac{1}{(n+2)!}$

15 DIFFERENCE EQUATIONS 2

Exercise 15A

1. 12, 29, 70

2. $p = 1$, $q = 6$

3. $+1$ for n even, -1 for n odd

4. Repeats after six terms

5. $\frac{1}{2}\left(1 + \sqrt{5}\right)$ (golden ratio)

6. 2

Exercise 15B

1. (a) $u_n = A3^n + B(-2)^n$

   (b) $u_n = A\left(2 + \sqrt{5}\right)^n + B\left(2 - \sqrt{5}\right)^n$

   (c) $u_n = A2^n + B(-1)^n$

2. $u_n = \dfrac{1}{\sqrt{5}}\left\{\left(\dfrac{1 + \sqrt{5}}{2}\right)^{n+1} - \left(\dfrac{1 - \sqrt{5}}{2}\right)^{n+1}\right\}$ $(n = 0, 1, \ldots)$

3. $u_n = \frac{1}{2}(1 - i)(2i)^n + \frac{1}{2}(1 + i)(-2i)^n$

4. $u_n = 4^n + 3 \times 2^n$; $u_6 = 4288$

5. $u_n = u_{n-1} + 6u_{n-2}$; $u_n = 3^n + 3 \times (-2)^n$

Activity 3

If $u_n =$ no. of pairs of mice; $u_n = u_{n-1} + 2u_{n-2}$

If $u_1 = u_2 = 10$, then $u_{12} = 13650$

Exercise 15C

1. (a) $u_n = A2^n + Bn2^n$

   (b) $u_n = A + Bn$

2. $u_n = 3^n(2 + n)$

3. $u_n = 2\left(i^n + (-i)^n\right)$

4. $u_n = (-1)^n(4 - 3n)$

5. (a) $u_n = 2 \times 5^n + 3 \times (-3)^n$

(b) $u_n = \frac{1}{2}\left(\left(\sqrt{3}\right)^n + \left(-\sqrt{3}\right)^n\right)$

(c) $u_n = 3^n(1+2n)$

6. $u_n = 2\left(u_{n-1} + u_{n-2}\right)$, $u_0 = u_1 = 1$;

$u_n = \frac{1}{2}\left(\left(1+\sqrt{3}\right)^n + \left(1-\sqrt{3}\right)^n\right)$

Activity 4

$u_n = -\frac{81}{4}\left(\frac{1}{3}\right)^n + 16\left(\frac{1}{2}\right)^n - \frac{3}{4} + \frac{1}{2}n$

Exercise 15D

1. $u_n = y_n + v_n$ where $y_n = A2^n + B3^n$ and

(a) $v_n = 1$

(b) $v_n = \frac{1}{4}(7+2n)$

(c) $v_n = \frac{1}{2}n^2 + \frac{7}{2}n + 8$

(d) $v_n = \frac{1}{6}\times 5^{n+2}$

(e) $v_n = -2n2^n$

2. $u_n = \frac{1}{2}\times 4^n - \frac{5}{3}\times 3^n + 2^{n+1}$

3. $u_n = \frac{1}{49}\left(29\times(-5)^{n-1} + 41\times 2^{n-1}\right) + \frac{2}{7}n2^n$

Exercise 15E

1. (a) $\dfrac{x}{1-2x-8x^2}$ (b) $\dfrac{2+7x}{1+x-3x^2}$

(c) $\dfrac{1+3x}{1-4x^2}$ (d) $\dfrac{1}{1-2x}$

2. (a) $\dfrac{1}{x-3} + \dfrac{2}{x+1}$ (b) $\dfrac{2}{2x-5} - \dfrac{1}{x-2}$

(c) $\dfrac{4}{x-3} - \dfrac{3}{x+3}$

3. (a) $1 + x + x^2 + \ldots + x^n + \ldots$

(b) $1 + 2x + 4x^2 + \ldots + 2^n x^n + \ldots$

(c) $1 - 3x + 9x^2 + \ldots + (-1)^n 3^n x^n + \ldots$

(d) $1 + 2x + 3x^2 + \ldots + (n+1)x^n + \ldots$

(e) $3\left(1 - 4x + 12x^2 + \ldots + (-1)^n(n+1)2^n x^n + \ldots\right)$

4. (a) $u_n = 4^{n+1} - 4(-1)^n$

(b) $u_n = 3\times 4^n$

5. $G(x) = \dfrac{1}{1-x-x^2}$

6. $u_n = 2\times 3^n + 3\times(-3)^n$

Exercise 15F

1. (a) $\dfrac{2}{1-2x} - \dfrac{1}{1-x}$ (b) $\dfrac{1}{3(2-x)} - \dfrac{5}{3(1+x)}$

(c) $\dfrac{1}{3(x+1)} + \dfrac{1}{6(1-2x)} + \dfrac{3}{2(1-2x)^2}$

2. (a) $\dfrac{1}{1-x}$ (b) $\dfrac{1}{(1-x)^2}$ (c) $\dfrac{-1}{x(1-x)^2} + \dfrac{1}{x}$

(d) $\dfrac{x^2}{(1-x)^2}$ (e) $\dfrac{1}{(1-2x)}$ (f) $\dfrac{25x^2}{(1-5x)}$

3. (a) $1 + 3x + 9x^2 + \ldots + 3^n x^n + \ldots$

(b) $\dfrac{1}{4} + \dfrac{x}{4} + \dfrac{3x^2}{16} + \ldots + \dfrac{(n+1)x^n}{2^{n+2}} + \ldots$

(c) $-x - x^2 - x^3 - \ldots + \left(-x^n\right) + \ldots$

4. (a) $u_n = \frac{1}{9}\left(7\times 4^n + (-2)^n - 8\right)$

(b) $u_n = 3^{n+1} - 2^{n+1}$

(c) $u_n = 4\times 2^n - \frac{1}{2}n^2 - \frac{5}{2}n - 4$

Miscellaneous Exercises

1. $u_n = 3\times 2^n + 2\times(-2)^n$

2. $u_n = (A + Bn)2^n$

3. Fibonacci Sequence

4. (a) $u_n = -1 + 3\times 2^{n-1}$

(b) $u_n = -n - 1 + 4\times 2^{n-1}$

(c) $u_n = \dfrac{1}{2\sqrt{3}}\left(\left(1+\sqrt{3}\right)^n - \left(1-\sqrt{3}\right)^n\right)$

5. $u_n = -\frac{2}{5}\times 2^n + \frac{3}{7}\times 4^n - \frac{1}{35}\times(-3)^n$

6. $N_t = 3 - \frac{27}{8}\left(\frac{4}{3}\right)^t + \frac{5}{4}\times 2^t$

7. $a = 2$, $b = -3$, $k = 9$; $u_n = \frac{1}{16}\left(-1 + (-3)^n\right) + \frac{9}{4}n$

8. $u_n = 5\left(3^n - 1\right)$; $n = 12$

9. $u_n = \frac{5}{8} + \frac{3}{8}\times(-1)^n + \frac{1}{4}n^2 + \frac{1}{2}n$

10. (a) (i) $(-3)^{n+1} + 1$ (ii) 1

(b) $u_n = 2u_{n-1} + 2u_{n-2}$;

initial condition $u_1 = 3$, $u_2 = 8$

(c) $u_n = 3^n$

11. $u_n = A3^n$; $\quad u_n = \frac{1}{2}\left(17 \times 3^{n-1} - 5\right)$

13. $u_n = 5 \times 2^n + (-1)^n$; $\quad u_n = 4 \times 2^n$; $\quad n = 8$

14. $u_n = A \times 4^n + B$

15. (a) $u_n = 3 \times (n-1)!$ $\qquad$ (b) $u_n = (A + Bn)(-2)^n$

16. $P_n = 4 + 2\left(-\frac{1}{2}\right)^n$; $\quad 4$

17. $u_n = \frac{1}{8} \times 4^n + \frac{3}{4} \times 2^n$; $\quad n = 8$

18. $u_n = \left(a + \frac{k}{(1-\alpha)}\right)(1+\alpha)^n - \frac{k2^n}{(1-\alpha)}$

(a) $n = 4$

(b) (ii) Population increases at first but will
eventually die out since the 2^n term will
dominate.

(iii) 7

FURTHER READING

As this is a new topic for schools and colleges, there are unfortunately few relevant resources. Most books on Discrete Mathematics (or Decision Mathematics) have been written for students taking courses in higher education. The list below gives the most relevant texts, but none of them cover all the material in the AEB syllabus at a suitable depth.

1. **Decision Mathematics,** 2nd edition, *The Spode Group* , 1986 (Ellis Horwood) 0 13 200973 0

2. **Decision Mathematics,** Vol. 1 and 2, *R. Davison* and *L. Cochrane,* 1991 (Cranfield) 1 871315 30 1, 1 871315 32 8

3. **Graphs, Networks and Design,** Open University Course TM361, 1981

4. **Essential Discrete Mathematics,** *R. Johnsonbaugh,* 1986 (Macmillan) 0 02 360630 4

5. **Discrete Mathematics,** *R. Johnsonbaugh,* 1984 (Macmillan) 0 02 360900 1

6. **Discrete Mathematics,** *Norman L. Biggs,* 1989 (OUP) 0 198 53252 0

7. **Modern Analytical Techniques,** 2nd edition, *F. Owen* and *R. Jones,* 1984 (Polytech) 0 85505 081 0

8. **Decision Making, Models and Algorithms,** *Saul I. Gass* 1985 (John Wiley & Sons) 0 471 80963 2

9. **Model Building in Mathematical Programming,** 2nd edition, *H. Williams,* 1984 (John Wiley & Sons) 0 471 90606 9

10. **Discrete Mathematics for New Technology,** *R. Garnier* and *J. Taylor,* 1992 (Adam Hilger) 0 7503 0136 8

INDEX